Partial differential equations arise in almost all areas of science, engineering, modeling, and forecasting. During the last two decades, pseudospectral methods have emerged as successful, and often superior, alternatives to better-known computational procedures - such as finite difference and finite element methods - in several key application areas. These areas include computational fluid dynamics, wave motion, and weather forecasting. This book explains how, when, and why this pseudospectral approach works. In order to make the subject accessible to students as well as to researchers and engineers, the presentation incorporates illustrations, examples, heuristic explanations, and algorithms rather than rigorous theoretical arguments. A key theme of the book is to establish and exploit the close connection that exists between pseudospectral and finite difference methods. This approach not only leads to new insights into already established pseudospectral procedures, but also provides many novel and powerful pseudospectral variations.

This book will be of interest to graduate students, scientists, and engineers interested in applying pseudospectral methods to real problems.

CAMBRIDGE MONOGRAPHS ON APPLIED AND COMPUTATIONAL MATHEMATICS

Series editors
P. G. CIARLET, A. ISERLES, R. V. KOHN, M. H. WRIGHT

1 A practical guide to pseudospectral methods

The *Cambridge Monographs on Applied and Computational Mathematics* reflects the crucial role of mathematical and computational techniques in contemporary science. The series publishes expositions on all aspects of applicable and numerical mathematics, with an emphasis on new developments in this fast-moving area of research.

State-of-the-art methods and algorithms as well as modern mathematical descriptions of physical and mechanical ideas are presented in a manner suited to graduate research students and professionals alike. Sound pedagogical presentation is a prerequisite. It is intended that books in the series will serve to inform a new generation of researchers.

A practical guide to pseudospectral methods

BENGT FORNBERG
University of Colorado

CAMBRIDGE UNIVERSITY PRESS
Cambridge, New York, Melbourne, Madrid, Cape Town, Singapore, São Paulo

Cambridge University Press
The Edinburgh Building, Cambridge CB2 2RU, UK

Published in the United States of America by Cambridge University Press, New York

www.cambridge.org
Information on this title: www.cambridge.org/9780521495820

© Cambridge University Press 1998

This publication is in copyright. Subject to statutory exception
and to the provisions of relevant collective licensing agreements,
no reproduction of any part may take place without
the written permission of Cambridge University Press.

First published 1995
First paperback edition 1998

A catalogue record for this publication is available from the British Library

ISBN-13 978-0-521-49582-0 hardback
ISBN-10 0-521-49582-2 hardback

ISBN-13 978-0-521-64564-5 paperback
ISBN-10 0-521-64564-6 paperback

Transferred to digital printing 2005

Contents

Preface

Partial differential equations arise in almost all areas of science, engineering, modeling, and forecasting. Finite difference and finite element methods have long histories as particularly flexible and powerful general-purpose numerical solution methods. In the last two decades, spectral and in particular pseudospectral (PS) methods have emerged as intriguing alternatives in many situations - and as superior ones in several areas.

The aim of this *Practical Guide* is to describe when, how, and why the PS approach works, in a style that makes the transition to actual numerical implementations as straightforward as possible. For this reason, the book focuses on illustrations, examples, and heuristic explanations, and includes key code segments and references, but contains only a few rigorous theorems or technical proofs. It is written primarily for scientists and engineers who are interested in applying PS methods to real problems. However, I also hope that it will prove suitable for graduate-level study, conveying to students that PS methods form an important and rapidly developing field in which elaborate mathematical preliminaries are unnecessary. Material that is normally included in undergraduate-level mathematics and numerical analysis courses is mentioned only if customary viewpoints need to be complemented.

The paper entitled "A Review of Pseudospectral Methods", which was co-authored by Professor David Sloan, appeared in *Acta Numerica 1994*. The encouragement offered by Dr. Arieh Iserles (the principal editor for *Acta Numerica*) was critical both for the original article and for its subsequent extension into this monograph. Many other colleagues have also helped with good advice and by bringing material to my attention. Special thanks go to Ben Herbst, Heinz-Otto Kreiss, and Lloyd N. Trefethen. Several ideas on both technical content and means of exposition have been "borrowed" from Professor Trefethen's (unfortunately unpublished) manuscript, "Finite Difference and Spectral Methods". This monograph was partly written while the author worked for Exxon Corporate Research.

My personal interest in PS methods goes back nearly to their historical origin. The work for my Ph.D. (in 1972 at Uppsala University, Sweden, under supervision of Professor H.-O. Kreiss) led to a lasting interest in this subject.

Boulder, Colorado *Bengt Fornberg*
September, 1995

List of abbreviations

AB	Adams–Bashforth
ADI	alternating direction implicit
AM	Adams–Moulton
BC	boundary condition
BDF	backward differentiation formula
BDF*n*	backward differentiation formula of order *n*
CF	cardinal function
CFL	Courant–Friedrichs–Lewy (stability condition)
Ch	Chebyshev
DCT	discrete convolution theorem
DFT	discrete Fourier transform
DM	differentiation matrix
DNS	direct numerical simulation
eq	equi-spaced
EV	eigenvalue
FCF	Fourier cardinal function
FCT	fast cosine transform
FD	finite differences
FE	finite elements
FFT	fast Fourier transform
FRFT	fractional Fourier transform
FST	fast sine transform
FV	finite volume
GQ	Gaussian quadrature
KdV	Korteweg–de Vreis (equation)
LAM	limited area model
Leg	Legendre (polynomial)
LES	large eddy simulation

LF	leapfrog
LHS	left-hand side
MOL	method of lines
NLS	nonlinear Schrödinger (equation)
NS	Navier–Stokes (equation)
ODE	ordinary differential equation
opt	optimal
PDE	partial differential equation
PS	pseudospectral
RHS	right-hand side
RK	Runge–Kutta method
RK*n*	Runge–Kutta method of order *n*
SOR	successive over-relaxation

1

Introduction

The partial differential equations (PDEs) that arise in applications can only rarely be solved in closed form. Even when they can be, the solutions are often impractical to work with and to visualize. Numerical techniques, on the other hand, can be applied successfully to virtually all well-posed PDEs. Broadly applicable techniques include finite element (FE), finite volume (FV), finite difference (FD), and, more recently, spectral methods. The complexity of the domain and the required levels of accuracy are often the key factors in selecting among these approaches.

Finite-element methods are particularly well suited to problems in very complex geometries (e.g. 3-D engineering structures), whereas spectral methods can offer superior accuracies (and cost efficiencies) mainly in simple geometries such as boxes and spheres (which can, however, be combined into more complex shapes). FD methods perform well over a broad range of accuracy requirements and (moderately complex) domains.

Both FE and FV methods are closely related to FD methods. FE methods can frequently be seen as a very convenient way to generate and administer complex FD schemes and to obtain results with relatively sharp error estimates. The connection between spectral methods - in particular the so-called pseudospectral (PS) methods, the topic of this book - and FD methods is closer still. A key theme in this book is to exploit this connection, both to make PS methods more intuitively understandable and to obtain particularly powerful and flexible PS variations.

Finite difference methods approximate derivatives of a function by *local* arguments (such as $du(x)/dx \approx [u(x+h)-u(x-h)]/2h$, where h is a small grid spacing; these methods are typically designed to be exact for polynomials of low order). This approach is very reasonable: because the derivative is a local property of a function (which need not be smooth), it seems unnecessary (and is certainly costly) to invoke many function values far away from the point of interest.

In contrast, spectral methods are *global*. A common way to introduce them starts by approximating the function we want to differentiate as a sum of very smooth basis functions:

$$u(x) \approx \sum_{k=0}^{N} a_k \phi_k(x),$$

where the $\phi_k(x)$ are for example Chebyshev polynomials or trigonometric functions. We then differentiate these functions exactly. In the context of solving time-dependent PDEs, this approach has notable strengths.

+ For analytic functions, errors typically decay (as N increases) at exponential rather than at (much slower) polynomial rates.
+ The method is virtually free of both dissipative and dispersive errors.

  In the context of solving high-Reynolds number fluid flows, the low physical dissipation will not be overwhelmed by large numerical dissipation. For convection-type problems, sharp gradients in a solution will not turn into wavetrains because of dispersive errors (making different frequency components propagate at different speeds).

+ The approach is surprisingly powerful for many cases in which both solutions and variable coefficients are nonsmooth or even discontinuous.
+ Especially in several space dimensions, the relatively coarse grids that suffice for most accuracy requirements allow very time- and memory-efficient calculations.

However, the following factors can cause difficulties or inefficiencies when using spectral methods:

- certain boundary conditions;
- irregular domains;
- strong shocks;
- variable resolution requirements in different parts of a large domain; and
- partly incomplete theoretical understanding.

In some applications – where these disadvantages are not present or can somehow be overcome – FE, FV, or FD methods do not even come close in efficiency. However, in most areas of application the situation is not so clear-cut. At present, spectral methods are highly successful in several areas: turbulence modeling, weather prediction, nonlinear waves, seismic modeling, et cetera; the list is growing (see e.g. Boyd 1989 for examples and references).

Spectral representations have been used for analytic studies of differential equations since the days of Fourier (1822). The idea of using them for numerical solutions of ordinary differential equations (ODEs) goes back at least to Lanczos (1938). Some present spectral methods can also be traced back to the "method of weighted residuals" (Finlayson and Scriven 1966). Their current popularity for PDEs dates back to the early 1970s. The major advance at that time was the pseudospectral approach of Kreiss and Oliger (1972), which - like most other spectral methods - benefited greatly from the fast Fourier transform (FFT) algorithm of Cooley and Tukey (1965).

Although PS methods are nowadays often introduced as indicated here (through expansions using smooth global functions, the topic of Chapter 2), there exists a very useful alternative: they can be seen as limiting cases of increasing-order FD methods. The basic idea (in the case of periodic problems) goes back to Kreiss and Oliger (1972), and was developed further by Fornberg (1975, 1987, 1990a,b). The introduction to Chapter 3 lists some of the advantages offered by this FD approach. In the remaining chapters, key properties and variations of PS methods are discussed, using whichever viewpoint is most illuminating. The last chapter briefly discusses some application areas.

2

Introduction to spectral methods via orthogonal functions

Spectral methods are usually described in the way we first indicated – as expansions based on global functions. Given a differential equation with boundary conditions, the idea is to approximate a solution $u(x)$ by a finite sum $v(x) = \sum_{k=0}^{N} a_k \phi_k(x)$; in the case of a time-dependent problem, $u(x,t)$ is approximated by $v(x,t)$ and $a_k(t)$. Two main questions arise: from which function class should $\phi_k(x)$, $k = 0, 1, \ldots$, be chosen; and how should the expansion coefficients a_k be determined. These questions are addressed in Sections 2.1 and 2.2. Section 2.3 introduces cardinal functions and differentiation matrices, important tools both for understanding and for computation. The last section describes Gibbs' phenomenon – the most notable example of how the expansion of a function loses accuracy in the vicinity of an irregularity.

Books that review this approach to spectral methods include Gottlieb and Orszag (1977), Voigt, Gottlieb, and Hussani (1984), Canuto et al. (1988), Boyd (1989), Mercier (1989), and Funaro (1992).

2.1. Function classes

Three requirements need to be met:

(1) the approximations $\sum_{k=0}^{N} a_k \phi_k(x)$ of $v(x)$ must converge rapidly (at least for reasonably smooth functions);

(2) given coefficients a_k, it should be easy to determine b_k such that

$$\frac{d}{dx}\left(\sum_{k=0}^{N} a_k \phi_k(x)\right) = \sum_{k=0}^{N} b_k \phi_k(x), \qquad (2.1\text{-}1)$$

and

(3) it should be fast to convert between coefficients a_k, $k = 0, \ldots, N$, and the values for the sum $v(x_i)$ at some set of nodes x_i, $i = 0, \ldots, N$.

Periodic problems. The choice here is easy: *trigonometric expansions* satisfy all the three requirements. The first two are immediate; the third became satisfied in 1965 through the FFT algorithm.

Nonperiodic problems. In this case, trigonometric expansions fail to satisfy requirement (1) – an irregularity will arise where the periodicity is artificially imposed. In case of a discontinuity, a Gibbs' phenomenon will occur (see Section 2.4). The coefficients a_k then decrease only like $O(1/N)$ as $N \to \infty$. Truncated *Taylor expansions* $v(x) = \sum_{k=0}^{N} a_k x^k$ will also fail on requirement (1), as convergence over $[-1, 1]$ requires extreme smoothness of $v(x)$ (analyticity throughout the unit circle).

The function class that has proven to be the most successful by far is *orthogonal polynomials* of Jacobi type, with Chebyshev and Legendre polynomials as the most important special cases. These polynomials are discussed in Appendix A. They arise in many contexts, such as the following.

- *Gaussian integration formulas* achieve a high accuracy by using zeros of orthogonal polynomials as nodes (see Section 4.7).
- *Singular Sturm–Liouville eigensystems* are well known to offer excellent bases for approximation. The Jacobi polynomials are the only polynomials that arise in this way.
- Truncated expansions in *Legendre polynomials* are optimal in the L^2 norm.

  For max-norm approximations of smooth functions, truncated *Chebyshev expansions* are particularly accurate.

- Interpolation at the *Chebyshev nodes* $x_k = -\cos(\pi k/N)$, $k = 0, 1, \ldots, N$, gives polynomials P_N^{Ch} that are always within a very small factor of the optimal polynomial in the max-norm approximation of any function $f(x)$:

$$\|f - P_N^{\text{Ch}}\| \le (1 + \Lambda_N^{\text{Ch}}) \|f - P_N^{\text{opt}}\|.$$

  Here, Λ_N^{Ch} is known as the Lebesgue constant of order N for Chebyshev interpolation. It is a particularly useful quantity since it depends only on N and not on the function that is interpolated. The way in which properties of f affect $\|f - P_N^{\text{opt}}\|$ is described by Jackson's theorems (Cheney 1966, Powell 1981). Λ_N^{Ch} is smaller than the corresponding constant for interpolation using Legendre nodes, and far superior to the disastrous one for equi-spaced interpolation:

$$\Lambda_N^{\text{Ch}} = O(\ln N);$$
$$\Lambda_N^{\text{Leg}} = O(\sqrt{N}); \tag{2.1-2}$$
$$\Lambda_N^{\text{eq}} = O\left(\frac{2^N}{N \ln N}\right).$$

These Lebesgue constants are discussed further in Section 3.3 and Appendix D.

The foregoing remarks confirm that Jacobi polynomials satisfy requirement (1) (to be discussed further in Sections 3.4 and 4.1). Because of the first-derivative recursions and the lack of explicit x-dependence therein (cf. bottom line in Table A-1), requirement (2) is met.

Applying the Chebyshev recursion to (2.1-1) and equating coefficients gives

$$\begin{bmatrix} 1 & 0 & -\frac{1}{2} & & & & & \\ & \frac{1}{4} & 0 & -\frac{1}{4} & & & & \\ & & \frac{1}{6} & 0 & -\frac{1}{6} & & & \\ & & & \frac{1}{8} & 0 & -\frac{1}{8} & & \\ & & & & \ddots & \ddots & \ddots & \\ & & & & & \frac{1}{2N-2} & 0 & \\ & & & & & & \frac{1}{2N} & \end{bmatrix} \times \begin{bmatrix} b_0 \\ b_1 \\ b_2 \\ b_3 \\ \vdots \\ b_{N-2} \\ b_{N-1} \end{bmatrix} = \begin{bmatrix} a_1 \\ a_2 \\ a_3 \\ a_4 \\ \vdots \\ a_{N-1} \\ a_N \end{bmatrix}. \tag{2.1-3}$$

Explicit formulas for $\{b_k\}$ in terms of $\{a_k\}$ are sometimes needed; see Appendix B.

Requirement (3) is satisfied for the Chebyshev case if we choose $x_i = -\cos(\pi i/N)$, $i = 0, \ldots, N$. Conversions between coefficients a_k and node values $v(x_i)$ can then be performed with a cosine version of the FFT algorithm (see Appendix F). As we will see in Sections 4.3 and 6.1 and in Figure F.3-1, the additional cost in the other cases need not be prohibitive. For these reasons, Chebyshev (and, to a lesser extent, Legendre) polynomials have become the almost universally preferred choice for nonperiodic spectral approximations.

Formulas similar to (2.1-3) can be found also when the LHS (left-hand side) of (2.1-1) is generalized to expressions involving rational coefficients and derivatives of many orders (Coutsias, Hagstrom, and Torres 1994):

$$\sum_{r=0}^{R} \frac{p_{m_r}(x)}{q_{n_r}(x)} \frac{d^r}{dx^r}\left(\sum_{k=0}^{N} a_k T_k(x)\right). \tag{2.1-4}$$

Here p_{m_r} and q_{n_r} are polynomials of degrees m_r and n_r respectively, and $r = 0, 1, 2, \ldots, R$. The coefficients $\{a_k\}$ and $\{b_k\}$ can be related through the following steps.

- Multiply by the denominators.
- Re-arrange the LHS into derivatives of products of polynomials by Leibniz's rule.
- Re-arrange all polynomials as expansions in Chebyshev polynomials.
- Express products of Chebyshev polynomials as simple Chebyshev polynomials through use of the relation $2T_m(x)T_n(x) = T_{m+n}(x)+T_{m-n}(x)$, $m \geq n$. If $T_m(x)\sum_{k=0}^{n} a_k T_k(x) = \sum_{k=0}^{m+n} b_k T_k(x)$, the matrix form of this becomes:

$$\begin{array}{lll} \text{row} & 1 & \Rightarrow \\ & \vdots & \\ \text{row} & m & \Rightarrow \\ & \vdots & \\ & & \\ & & \\ & \vdots & \\ \text{row} & m+n & \Rightarrow \end{array} \begin{bmatrix} & & 1 & & & \\ & 1 & & 1 & & \\ 2 & & & & \ddots & \\ & 1 & & & & 1 \\ & & 1 & & & \\ & & & 1 & & \\ & & & & \ddots & \\ & & & & & 1 \end{bmatrix} \begin{bmatrix} a_0 \\ a_1 \\ \vdots \\ \vdots \\ a_n \end{bmatrix} = 2 \begin{bmatrix} b_0 \\ b_1 \\ \vdots \\ \\ \\ \vdots \\ b_{m+n} \end{bmatrix},$$

  with more diagonals present in cases of more complex multiplying functions.
- Use (repeatedly) (2.1-3).

A few special cases of (2.1-4) are addressed in the appendix of Gottlieb and Orszag (1977). Relations between expansion coefficients such as these are essential for Galerkin and tau approximations, but not for collocation (PS) approximations; see Section 2.2 and Appendix B.

2.2. Techniques for determining expansion coefficients

The three main techniques used to determine the expansion coefficients a_k are the tau, Galerkin, and collocation (PS) methods. In all cases, we consider the residual $R_n(x)$ (or $R_n(x,t)$) when an expansion is substituted into the governing equation. We want to keep the residual as small as possible across the domain while satisfying the boundary conditions.

Tau. Require that a_k be selected so that the boundary conditions are satisfied, and make the residual *orthogonal* to as many of the basis functions as possible.

Galerkin. Combine the original basis functions into a new set in which all the functions satisfy the boundary conditions. Then require that the residual be orthogonal to as many of these new basis functions as possible.

Collocation (PS). This is similar to the tau method: Require that a_k be selected so that the boundary conditions are satisfied, but make the residual *zero* at as many (suitably chosen) spatial points as possible.

Implementation details for a linear, time-independent model problem are given in Appendix B.

For linear problems, the systems of equations that arise from the tau and Galerkin methods can sometimes be solved rapidly thanks to favorable sparsity patterns. However, variable coefficients and nonlinearities are often difficult to handle. The tau method was first used by Lanczos (1938). The Galerkin procedure is central to FE methods; spectral (global) versions of it have been in use since the mid-1950s. The FFT algorithm and contributions by Orszag (1969 and 1970, on ways to deal with nonlinearities) encouraged the use of these methods.

As illustrated by the example in Appendix B, the collocation (PS) method can be viewed as a method of finding numerical approximations to derivatives at gridpoints. Then, in a finite difference–like manner, the governing equations are satisfied pointwise in physical space. The PS method then becomes particularly easy to apply to equations with variable coefficients and nonlinearities, since these give rise only to products of numbers (rather than to problems of determining the expansion coefficients for products of expansions). This is how the collocation approach was originally presented (for PDEs with periodic solutions) by Kreiss and Oliger (1972). It was referred to as the *pseudospectral* method in Orszag (1972).

The rest of this book will focus on the PS method.

2.3. Cardinal functions: example of a differentiation matrix

The concepts of cardinal functions (CFs) and differentiation matrices (DMs) are both theoretically and numerically useful well beyond the realm of methods derived from orthogonal functions. Therefore, we postpone the main discussion of these until Sections 4.3 and 4.4 (when our background is more general) and consider them here only in the case of the Fourier–PS method.

The interpolating trigonometric polynomial to periodic data can be thought of as a weighted sum of CFs, each with the property of having unit value at one of the data points and zero at the rest. This is much like how Lagrange's interpolation formula works, with the main difference that, in the Fourier case, all CFs are simply translates of each other.

Assume for simplicity that we have an odd number $N=2m+1$ of gridpoints at locations $x_i = i/(m+\frac{1}{2})$, $i=-m,\dots,-1,0,1,\dots,m$ in $[-1,1]$. By inspection,

$$\phi_m(x) = \frac{2}{N}\left\{\frac{1}{2} + \cos\pi x + \cos 2\pi x + \cdots + \cos m\pi x\right\}$$
$$= \frac{\sin(N\pi x/2)}{N\sin(\pi x/2)} \tag{2.3-1}$$

is a $[-1,1]$-periodic trigonometric polynomial that satisfies

$$\phi_m(x_i) = \begin{cases} 1 & \text{if } i = 0 \quad [\pm N, \pm 2N, \ldots \text{ if periodically extended}], \\ 0 & \text{otherwise.} \end{cases}$$

Figure 2.3-1 displays the CF $\phi_8(x)$, shows how translates of it add up to give the trigonometric interpolant to a step function, and graphs the Gibbs' phenomenon - a finite "overshoot" beside a discontinuity.

The cardinal function

$$\operatorname{sinc}(x) = \frac{\sin\pi x}{\pi x} = \lim_{m\to\infty} \phi_m\left(\frac{x}{m}\right) \tag{2.3-2}$$

is often convenient to use in analysis. Sir E. T. Whittaker (1915) found this CF quite noteworthy: "... a function of royal blood ... whose distinguished properties separates it from its bourgeois brethren". Additional references on CFs can be found e.g. in J. M. Whittaker (1927) and Stenger (1993).

From (2.3-1) follows

$$\left.\frac{d}{dx}\phi_m(x)\right|_{x=x_i} = \begin{cases} 0 & \text{if } i = 0, \\ \dfrac{(-1)^i\pi}{2\sin(i\pi/N)} & \text{otherwise.} \end{cases} \tag{2.3-3}$$

Given a vector of periodic data values $v(x_j)$, $j = -m, \ldots, m$, the interpolating polynomial can be written

$$v(x) = \sum_{j=-m}^{m} v(x_j)\phi_m(x - x_j).$$

The derivative of this function at $x = x_i$ becomes

$$v'(x_i) = \sum_{j=-m}^{m} v(x_j)\left.\frac{d}{dx}\phi_m(x - x_j)\right|_{x=x_i} = \sum_{j=-m}^{m} v(x_j)\left.\frac{d}{dx}\phi_m(x)\right|_{x=x_{i-j}}.$$

Written in matrix form, this becomes

$$\begin{bmatrix} v'(x_{-m}) \\ \vdots \\ \vdots \\ \vdots \\ v'(x_m) \end{bmatrix} = \begin{bmatrix} & & \\ & D & \\ & & \end{bmatrix} \begin{bmatrix} v(x_{-m}) \\ \vdots \\ \vdots \\ \vdots \\ v(x_m) \end{bmatrix},$$

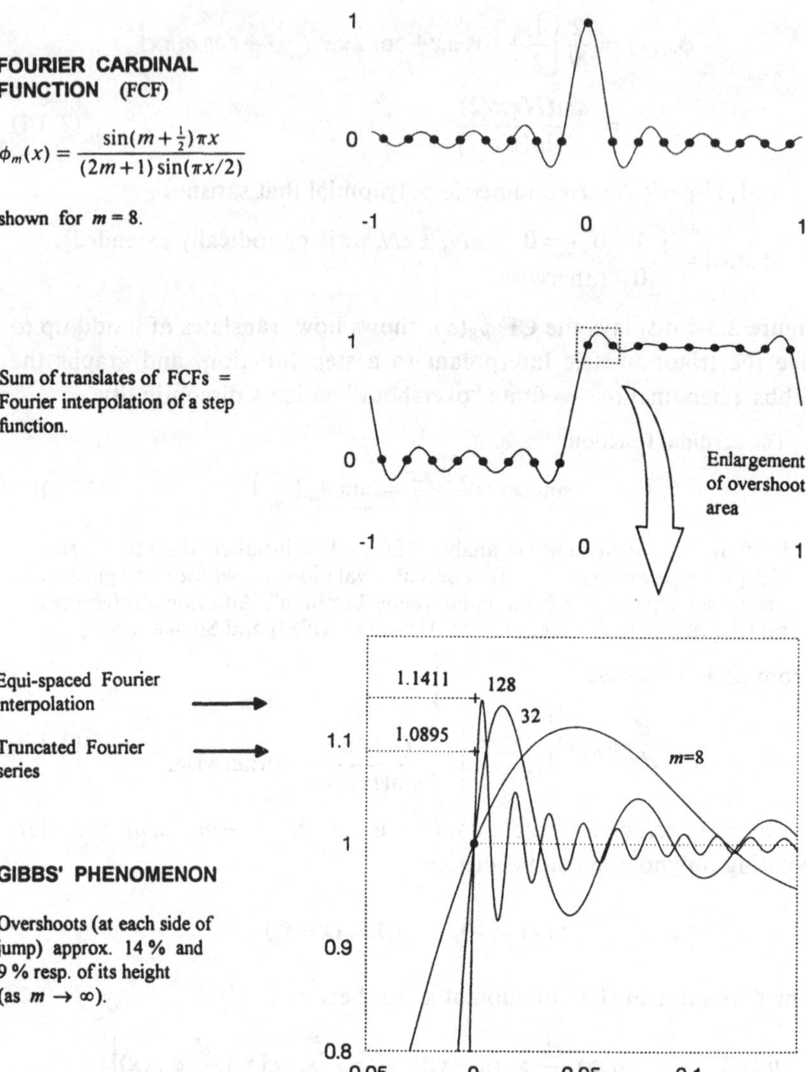

Figure 2.3-1. Fourier cardinal functions and Gibbs' phenomenon.

where the differentiation matrix D has the elements (making use of equation (2.2-3)):

$$D_{i,j} = \frac{d}{dx}\phi_m(x)\bigg|_{x=x_{i-j}} = \begin{cases} \dfrac{\pi(-1)^{i-j}}{2\sin(\pi(i-j)/N)} & \text{if } i \neq j, \\ 0 & \text{if } i = j. \end{cases} \tag{2.3-4}$$

2.4. Gibbs' phenomenon

The overshoot shown in the lower portion of Figure 2.3-1 arises whenever a discontinuous function is expanded in or interpolated with smooth functions.

What we now call Gibbs' phenomenon was first noted by Wilbraham (1848). Unaware of this, Michelson and Stratten (1898) found traces of these overshoots in the output plots from a mechanical Fourier analyzer they had constructed.

Michelson and Stratten's analyzer is described in some detail under the entry "Calculating machines" in the 1910 (11th) edition of The Encyclopedia Britannica. Some hardware is preserved (but not normally on display) at the Smithsonian in Washington, DC. The analyzer was a refinement (and extension to about 80 modes) of an earlier version invented by Lord Kelvin for the calculation of tides. Kelvin's device was so well suited for its task that it remained in use 20 years into the era of electronic computers.

This observation by Michelson (who is probably best known for his ether experiment with Morley) prompted him to write a letter to *Nature* inquiring about the convergence properties of a Fourier series for a discontinuous function. In reply, J. Gibbs (an eminent chemist) provided first a flawed and then a satisfactory answer. Some historical notes on Gibbs' phenomenon can be found in Hewitt and Hewitt (1979).

Two different variations of Gibbs' phenomenon arise in spectral methods. The overshoots on a jump of height 1 become as follows.

Equi-spaced Fourier interpolation. The notation will be simpler if we first transform to an infinite interval. Following the line of reasoning indicated in Figure 2.3-1 and noting equation (2.3-2), we have:

$$\begin{aligned} G_I &= \max_{0<\xi<1}\left\{\frac{\sin \pi\xi}{\pi\xi} + \frac{\sin \pi(\xi-1)}{\pi(\xi-1)} + \frac{\sin \pi(\xi-2)}{\pi(\xi-2)} + \cdots\right\} - 1 \\ &= \max_{0<\xi<1}\left\{\frac{\sin \pi\xi}{\pi} \sum_{k=0}^{\infty} \frac{(-1)^k}{\xi-k}\right\} - 1 \approx 0.1411. \end{aligned}$$

Truncated Fourier expansion. We consider again a piecewise constant function with a unit jump at the origin:

$$f(x) = \begin{cases} \frac{1}{2} & \text{if } 0 < x < \pi, \\ -\frac{1}{2} & \text{if } -\pi < x < 0. \end{cases}$$

The Fourier series (of a 2π-periodic extension) of this function is

$$f(x)=\frac{2}{\pi}\sum_{k=0}^{\infty}\frac{\sin(2k+1)x}{2k+1}.$$

The derivative of its truncated sum

$$f_N(x)=\frac{2}{\pi}\sum_{k=0}^{N}\frac{\sin(2k+1)x}{2k+1}$$

is

$$f_N'(x)=\frac{2}{\pi}\sum_{k=0}^{N}\cos(2k+1)x=\frac{\sin 2(N+1)x}{\pi\sin x};$$

that is, the extrema of $f_N(x)$ for $x>0$ occur at $x_{N,j}=\pi j/2(N+1)$, $j=1,2,\dots$. At these points,

$$f_N(x_{N,j})=\frac{1}{\pi(N+1)}\sum_{k=0}^{N}\frac{\sin((2k+1)\pi j/2(N+1))}{(2k+1)/2(N+1)}.$$

The sum is a discrete approximation to the integral

$$\frac{1}{\pi}\int_0^1\frac{\sin j\pi t}{t}\,dt=\frac{1}{\pi}\int_0^{j\pi}\frac{\sin t}{t}\,dt.$$

The maximum value is taken for $j=1$, yielding $(1/\pi)\int_0^{\pi}((\sin t)/t)\,dt\approx 0.5895$. Thus, as $N\to\infty$, the overshoot of a unit-height jump approaches $G_T\approx 0.0895$.

Gibbs' phenomenon for Chebyshev (Jacobi) expansions is essentially the same as in the Fourier case when the irregularities are located inside $[-1,1]$. However, the interpolation overshoot from a unit jump at $x=\pm 1$ is larger:

$$G_{I,\pm 1}=\max_{1\le\xi\le 2}\left\{-\frac{\sin\pi\xi}{\pi\xi}\right\}\approx 0.2172.$$

Gibbs' phenomenon (the $O(1)$ error next to a discontinuity) is the most notable instance of how an irregularity of a piecewise smooth function can affect the convergence of both interpolants and truncated series expansions; see Table 2.4-1. The decay rates of Fourier expansion coefficients are the same as the order of the maximum norm of errors away from irregularities.

For continuous but not piecewise differentiable functions, a Fourier series can do such strange things as diverging to infinity at some point(s) (in spite of each truncation giving the best possible least-squares approximation to the function when using up to that number of terms!). Such subtleties have no numerical consequences - Table 2.4-1 is a good guide for all situations of numerical relevance.

Table 2.4-1. *Order of max-norm errors caused by irregularities of a function*

Function	Max-norm of errors (order)	
	Near irregularity	Away from irregularity
f discontinuous	1	$1/N$
f' discontinuous	$1/N$	$1/N^2$
f'' discontinuous	$1/N^2$	$1/N^3$
⋮	⋮	⋮
f analytic (periodic)	e^{-cN}, $c>0$	

3

Introduction to PS methods via finite differences

Both periodic and nonperiodic PS methods can be seen as high-accuracy limits of FD methods. This alternative approach to PS methods provides both generalizations and insights.

- Orthogonal polynomials and functions lead only to a small class of possible spectral methods, whereas the FD viewpoint allows many generalizations.

  For example, all classical orthogonal polynomials cluster the nodes quadratically toward the ends of the interval - this is often, but not always, best.

- An FD viewpoint offers a chance to explore intermediate methods between low-order FD and PS methods.

  One might consider procedures of not quite as high order as Chebyshev spectral methods, and with nodes not clustered quite as densely - possibly trading some accuracy for stability and simplicity.

- Two separate ways to view any method will always provide more opportunities for analysis, understanding, and improvement.
- Many special enhancements have been designed for FD methods. Viewing PS methods as a special case of FD methods often makes it easier to carry over such ideas.

  Examples include staggered grids, upwind techniques, boundary techniques, polar and spherical coordinates, etc. (see Chapters 5 and 6).

- Comparisons between PS and FD methods can be made more consistent.

Sections 3.1 and 3.2 contain some general material on FD approximations, allowing us in Section 3.3 to discuss different types of node distributions. The relation between these types and the accuracy that polynomial interpolation provides (at different locations over $[-1, 1]$) is clarified in Section 3.4. In Section 3.5, we consider the limiting FD method of formally infinite order for periodic, equi-spaced data. The DM turns out to be identical to that obtained for the Fourier–PS method in Section 2.3. Hence, the two methods are identical. This equivalence is extended and discussed further in Section 3.6.

Although some of the earliest PS studies (including the original 1972 paper on the periodic PS method by Kreiss and Oliger) implied a close connection between FD and PS methods, Fornberg and Sloan (1994) – and this volume – are apparently the first review works that exploit this FD–PS connection in a systematic way.

3.1. Algorithm for finding FD weights on arbitrarily spaced grids

Centered FD formulas for equi-spaced grids are readily available from tables, and can be derived by symbolic manipulation of difference operators. For example, the centered approximations (at a gridpoint) for the first derivative

$$\begin{aligned} f'(x) &= [\qquad -\tfrac{1}{2}f(x-h)+0f(x)+\tfrac{1}{2}f(x+h) \qquad]/h \\ &\quad +O(h^2) \\ &= [\tfrac{1}{12}f(x-2h)-\tfrac{2}{3}f(x-h)+0f(x)+\tfrac{2}{3}f(x+h)-\tfrac{1}{12}f(x+2h)]/h \\ &\quad +O(h^4) \\ &\ \ \vdots \end{aligned}$$

are exact for all polynomials of (respectively) degrees 2, 4, ... It is convenient to collect weights in tables such as Table 3.1-1 (for the kth derivative, divide by h^k).

Another equi-spaced case of interest is one-sided stencils, whose use is often necessary at boundaries. Table 3.1-2 shows some weights for this case.

The one-sided weights for the first derivative are the same ones that arise in backward differentiation formulas (BDFs) for ODEs; cf. equation (3.2-3), Section G.2(c), and, for a more general discussion, Lambert (1991).

To explore general properties of FD schemes (and, more importantly, to use such schemes), it is desirable to have a simple algorithm for the following problem.

Given: Gridpoints $x_0, x_1, \ldots, x_n$ (nonrepeated, but otherwise arbitrary); point $x=\xi$ at which the approximations are wanted (may, but need not be at a gridpoint); and m, the highest order of derivative that is of interest.

Table 3.1-1. *Weights for some centered FD formulas on an equi-spaced grid*

Order of accuracy	Approximations at $x=0$; x coordinates at nodes: -4	-3	-2	-1	0	1	2	3	4
0th derivative									
∞					1				
1st derivative									
2				$-\frac{1}{2}$	0	$\frac{1}{2}$			
4			$\frac{1}{12}$	$-\frac{2}{3}$	0	$\frac{2}{3}$	$-\frac{1}{12}$		
6		$-\frac{1}{60}$	$\frac{3}{20}$	$-\frac{3}{4}$	0	$\frac{3}{4}$	$-\frac{3}{20}$	$\frac{1}{60}$	
8	$\frac{1}{280}$	$-\frac{4}{105}$	$\frac{1}{5}$	$-\frac{4}{5}$	0	$\frac{4}{5}$	$-\frac{1}{5}$	$\frac{4}{105}$	$-\frac{1}{280}$
2nd derivative									
2				1	-2	1			
4			$-\frac{1}{12}$	$\frac{4}{3}$	$-\frac{5}{2}$	$\frac{4}{3}$	$-\frac{1}{12}$		
6		$\frac{1}{90}$	$-\frac{3}{20}$	$\frac{3}{2}$	$-\frac{49}{18}$	$\frac{3}{2}$	$-\frac{3}{20}$	$\frac{1}{90}$	
8	$-\frac{1}{560}$	$\frac{8}{315}$	$-\frac{1}{5}$	$\frac{8}{5}$	$-\frac{205}{72}$	$\frac{8}{5}$	$-\frac{1}{5}$	$\frac{8}{315}$	$-\frac{1}{560}$
3rd derivative									
2			$-\frac{1}{2}$	1	0	-1	$\frac{1}{2}$		
4		$\frac{1}{8}$	-1	$\frac{13}{8}$	0	$-\frac{13}{8}$	1	$-\frac{1}{8}$	
6	$-\frac{7}{240}$	$\frac{3}{10}$	$-\frac{169}{120}$	$\frac{61}{30}$	0	$-\frac{61}{30}$	$\frac{169}{120}$	$-\frac{3}{10}$	$\frac{7}{240}$
4th derivative									
2			1	-4	6	-4	1		
4		$-\frac{1}{6}$	2	$-\frac{13}{2}$	$\frac{28}{3}$	$-\frac{13}{2}$	2	$-\frac{1}{6}$	
6	$\frac{7}{240}$	$-\frac{2}{5}$	$\frac{169}{60}$	$-\frac{122}{15}$	$\frac{91}{8}$	$-\frac{122}{15}$	$\frac{169}{60}$	$-\frac{2}{5}$	$\frac{7}{240}$

Find: Weights $c_{i,j}^k$ such that the approximations

$$\left.\frac{d^k f}{dx^k}\right|_{x=\xi} \approx \sum_{j=0}^{i} c_{i,j}^k f(x_j), \quad k=0,1,\ldots,m, \quad i=k,k+1,\ldots,n \qquad (3.1\text{-}1)$$

are all optimal, that is, are exact for all polynomials of as high degree as possible (to interpret the indices on the weights $c_{i,j}^k$, see Table 3.1-3).

A short and fast algorithm for this problem was discovered only recently (Fornberg 1988b; in more detail, 1992) as follows.

Table 3.1-2. *Weights for some one-sided FD formulas on an equi-spaced grid*

Order of accuracy	Approximations at $x = 0$; x coordinates at nodes: 0	1	2	3	4	5	6	7	8
0th derivative									
∞	1								
1st derivative									
1	-1	1							
2	$-\frac{3}{2}$	2	$-\frac{1}{2}$						
3	$-\frac{11}{6}$	3	$-\frac{3}{2}$	$\frac{1}{3}$					
4	$-\frac{25}{12}$	4	-3	$\frac{4}{3}$	$-\frac{1}{4}$				
5	$-\frac{137}{60}$	5	-5	$\frac{10}{3}$	$-\frac{5}{4}$	$\frac{1}{5}$			
6	$-\frac{49}{20}$	6	$-\frac{15}{2}$	$\frac{20}{3}$	$-\frac{15}{4}$	$\frac{6}{5}$	$-\frac{1}{6}$		
7	$-\frac{363}{140}$	7	$-\frac{21}{2}$	$\frac{35}{3}$	$-\frac{35}{4}$	$\frac{21}{5}$	$-\frac{7}{6}$	$\frac{1}{7}$	
8	$-\frac{761}{280}$	8	-14	$\frac{56}{3}$	$-\frac{35}{2}$	$\frac{56}{5}$	$-\frac{14}{3}$	$\frac{8}{7}$	$-\frac{1}{8}$
2nd derivative									
1	1	-2	1						
2	2	-5	4	-1					
3	$\frac{35}{12}$	$-\frac{26}{3}$	$\frac{19}{2}$	$-\frac{14}{3}$	$\frac{11}{12}$				
4	$\frac{15}{4}$	$-\frac{77}{6}$	$\frac{107}{6}$	-13	$\frac{61}{12}$	$-\frac{5}{6}$			
5	$\frac{203}{45}$	$-\frac{87}{5}$	$\frac{117}{4}$	$-\frac{254}{9}$	$\frac{33}{2}$	$-\frac{27}{5}$	$\frac{137}{180}$		
6	$\frac{469}{90}$	$-\frac{223}{10}$	$\frac{879}{20}$	$-\frac{949}{18}$	41	$-\frac{201}{10}$	$\frac{1019}{180}$	$-\frac{7}{10}$	
7	$\frac{29531}{5040}$	$-\frac{962}{35}$	$\frac{621}{10}$	$-\frac{4006}{45}$	$\frac{691}{8}$	$-\frac{282}{5}$	$\frac{2143}{90}$	$-\frac{206}{35}$	$\frac{363}{560}$
3rd derivative									
1	-1	3	-3	1					
2	$-\frac{5}{2}$	9	-12	7	$-\frac{3}{2}$				
3	$-\frac{17}{4}$	$\frac{71}{4}$	$-\frac{59}{2}$	$\frac{49}{2}$	$-\frac{41}{4}$	$\frac{7}{4}$			
4	$-\frac{49}{8}$	29	$-\frac{461}{8}$	62	$-\frac{307}{8}$	13	$-\frac{15}{8}$		
5	$-\frac{967}{120}$	$\frac{638}{15}$	$-\frac{3929}{40}$	$\frac{389}{3}$	$-\frac{2545}{24}$	$\frac{268}{5}$	$-\frac{1849}{120}$	$\frac{29}{15}$	
6	$-\frac{801}{80}$	$\frac{349}{6}$	$-\frac{18353}{120}$	$\frac{2391}{10}$	$-\frac{1457}{6}$	$\frac{4891}{30}$	$-\frac{561}{8}$	$\frac{527}{30}$	$-\frac{469}{240}$
4th derivative									
1	1	-4	6	-4	1				
2	3	-14	26	-24	11	-2			
3	$\frac{35}{6}$	-31	$\frac{137}{2}$	$-\frac{242}{3}$	$\frac{107}{2}$	-19	$\frac{17}{6}$		
4	$\frac{28}{3}$	$-\frac{111}{2}$	142	$-\frac{1219}{6}$	176	$-\frac{185}{2}$	$\frac{82}{3}$	$-\frac{7}{2}$	
5	$\frac{1069}{80}$	$-\frac{1316}{15}$	$\frac{15289}{60}$	$-\frac{2144}{5}$	$\frac{10993}{24}$	$-\frac{4772}{15}$	$\frac{2803}{20}$	$-\frac{536}{15}$	$\frac{967}{240}$

Table 3.1-3. *Schematic illustration of the notation used for weights generated by the fast algorithm*

Order of accuracy	Approximations at $x=\xi$; x coordinates at nodes:						
	x_0	x_1	x_2	x_3	......		x_n
0th derivative							
1	$c^0_{0,0}$						
2	$c^0_{1,0}$	$c^0_{1,1}$					
3	$c^0_{2,0}$	$c^0_{2,1}$	$c^0_{2,2}$				
4	$c^0_{3,0}$	$c^0_{3,1}$	$c^0_{3,2}$	$c^0_{3,3}$	⋱		
⋮	⋮	⋮	⋮	⋮			
$n+1$	$c^0_{n,0}$	$c^0_{n,1}$	$c^0_{n,2}$	$c^0_{n,3}$	......		$c^0_{n,n}$
1st derivative							
1	$c^1_{1,0}$	$c^1_{1,1}$					
2	$c^1_{2,0}$	$c^1_{2,1}$	$c^1_{2,2}$				
3	$c^1_{3,0}$	$c^1_{3,1}$	$c^1_{3,2}$	$c^1_{3,3}$	⋱		
⋮	⋮	⋮	⋮	⋮			
n	$c^1_{n,0}$	$c^1_{n,1}$	$c^1_{n,2}$	$c^1_{n,3}$	......		$c^1_{n,n}$
⋮							
mth derivative							
1	$c^m_{m,0}$	$c^m_{m,1}$	$c^m_{m,2}$	$c^m_{m,3}$	... $c^m_{m,m}$		
2	$c^m_{m+1,0}$	$c^m_{m+1,1}$	$c^m_{m+1,2}$	$c^m_{m+1,3}$	... $c^m_{m+1,m}$	$c^m_{m+1,m+1}$	
⋮	⋮	⋮	⋮	⋮	⋮	⋮	⋱
$n-m+1$	$c^m_{n,0}$	$c^m_{n,1}$	$c^m_{n,2}$	$c^m_{n,3}$	......		$c^m_{n,n}$

$c^0_{0,0} := 1,\ \alpha := 1$

for $i := 1$ to n

  $\beta := 1$

  for $j := 0$ to $i-1$

    $\beta := \beta(x_i - x_j)$

    for $k := 0$ to $\min(i,m)$

      $c^k_{i,j} := ((x_i-\xi)c^k_{i-1,j} - kc^{k-1}_{i-1,j})/(x_i-x_j)$

  for $k := 0$ to $\min(i,m)$

    $c^k_{i,i} := \alpha(kc^{k-1}_{i-1,i-1} - (x_{i-1}-\xi)c^k_{i-1,i-1})/\beta$

  $\alpha := \beta.$

Notes

(1) Any non-initialized quantity referred to is assumed to be equal to zero.

(2) Only four operations are needed for each weight (to leading order; note that the subtractions $x_i - \xi$ and $x_i - x_j$ can be moved out of the innermost loop).

(3) The calculation of weights is numerically stable. However, especially in case of high derivatives, *applying* FD weights to a function can be numerically ill-conditioned and can lead to severe cancellations and loss of significant digits.

For analytic functions that can be evaluated in the complex plane, finite difference formulas can be very accurate and quite well-conditioned also for high derivatives. Fornberg (1981a,b) gives a routine that automatically selects (complex) points and order of approximation for best accuracy. For analytic functions that satisfy simple ODEs (thus excluding important classes of functions such as those involving $\Gamma(x)$, $\zeta(x)$, etc.), high derivatives can often be calculated recursively (Barton, Willers, and Zahar 1971, Corliss and Chang 1982).

(4) The case $m = 0$ offers a very fast way for polynomial interpolation at a single point (e.g. compared to Aitken's and Neville's algorithms; cf. Henrici 1964). The ability to re-use weights can offer further savings (e.g. on 2-D and 3-D grids).

(5) If we are interested in the weights for only the stencils based on all the gridpoints x_j, $j = 0, 1, \ldots, n$ (and not in the lower-order stencils based on fewer points), then we can omit the first of the two subscripts for c (i.e., it suffices to declare a 2-D array to hold $c_{0\ \text{to}\ n}^{0\ \text{to}\ m}$). A slight re-ordering of the arithmetic operations will allow safe overwriting within this array.

FORTRAN codes (and a test driver) are given in Appendix C.

The derivation of the algorithm starts by noting that the best polynomial-based approximation to d^kf/dx^k at some point ξ is obtained by differentiating Lagrange's interpolating polynomial (to be described shortly; cf. equation (3.3-1)):

$$\frac{d^k p_N(x)}{dx^k} = \sum_{j=0}^{N} \frac{d^k F_j(x)}{dx^k} \cdot f(x_j).$$

Thus

$$\left.\frac{d^k p_N(x)}{dx^k}\right|_{x=\xi} = \sum_{j=0}^{N} \left.\frac{d^k F_j(x)}{dx^k}\right|_{x=\xi} \cdot f(x_j). \tag{3.1-2}$$

Considering not only interpolations based on all gridpoints x_j, $j = 0, 1, \ldots, N$, but also interpolations based on leading subsets $\{x_j, j = 0, 1, \ldots, i\}$, $i = 0, 1, \ldots, N$, and furthermore simplifying the notation by assuming $\xi = 0$, we write (3.1-2) as

$$\left.\frac{d^k p_i(x)}{dx^k}\right|_{x=0} = \sum_{j=0}^{i} \left.\frac{d^k F_{i,j}(x)}{dx^k}\right|_{x=0} \cdot f(x_j), \quad i = 0, 1, \ldots, N.$$

The desired weights are therefore

$$c_{i,j}^k = \left.\frac{d^k F_{i,j}(x)}{dx^k}\right|_{x=0}.$$

By Taylor's theorem, we can write $F_{i,j}(x)$ as

$$F_{i,j}(x) = \sum_{k=0}^{i} \frac{c_{i,j}^k}{k!} x^k. \tag{3.1-3}$$

The problem of determining the weights $c_{i,j}^k$ is therefore equivalent to that of re-arranging

$$F_{i,j}(x) = \frac{(x-x_0)(x-x_1)\cdots(x-x_{j-1})(x-x_{j+1})\cdots(x-x_i)}{(x_j-x_0)(x_j-x_1)\cdots(x_j-x_{j-1})(x_j-x_{j+1})\cdots(x_j-x_i)} \tag{3.1-4}$$

(cf. equation (3.3-2)) into powers of x. A convenient way to achieve this re-arranging starts by noting that (3.1-4) implies a couple of recursion relations:

$$F_{i,j}(x) = \frac{x-x_i}{x_j-x_i} F_{i-1,j}(x),$$

$$F_{i,i}(x) = \frac{\prod_{l=0}^{i-2}(x_{i-1}-x_l)}{\prod_{l=0}^{i-1}(x_i-x_l)} (x-x_{i-1}) F_{i-1,i-1}(x).$$

Substituting (3.1-3) into these relations gives, after equating coefficients, two recursions for the weights $c_{i,j}^k$. Starting with $c_{0,0}^0 = 1$, these recursions suffice to determine all $c_{i,j}^k$, $k = 0, 1, \ldots, m$, $j = k, k+1, \ldots, N$, in a fast and numerically stable manner.

3.2. Growth rates of FD weights on equi-spaced grids

Figures 3.2-1 and 3.2-2 illustrate how the magnitudes of the weights for the first derivative grow with increasing orders of accuracy (cf. Tables 3.1-1 and 3.1-2).

In the centered case, approximations of increasing orders of accuracy converge to a limit method of formally infinite order. For the first derivative ($m = 1$), the limit can be found directly from the closed-form expression for the weights

$$c_{p,j}^1 = \begin{cases} \dfrac{(-1)^{j+1}(p/2)!^2}{j(p/2+j)!\,(p/2-j)!} & \text{if } j = \pm 1, \pm 2, \ldots, \pm p/2, \\ 0 & \text{if } j = 0, \end{cases} \tag{3.2-1}$$

where p (even) is the approximation's order of accuracy and j the x position of the weight. In the limit of $p \to \infty$,

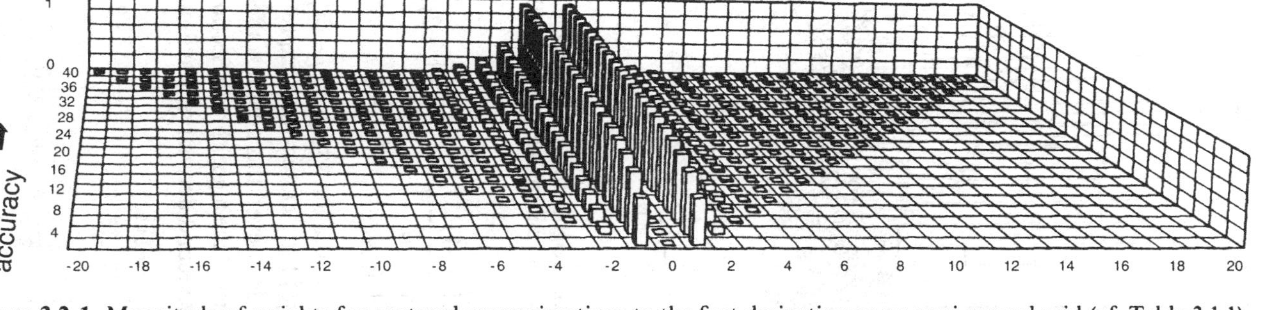

Figure 3.2-1. Magnitude of weights for centered approximations to the first derivative on an equi-spaced grid (cf. Table 3.1-1).

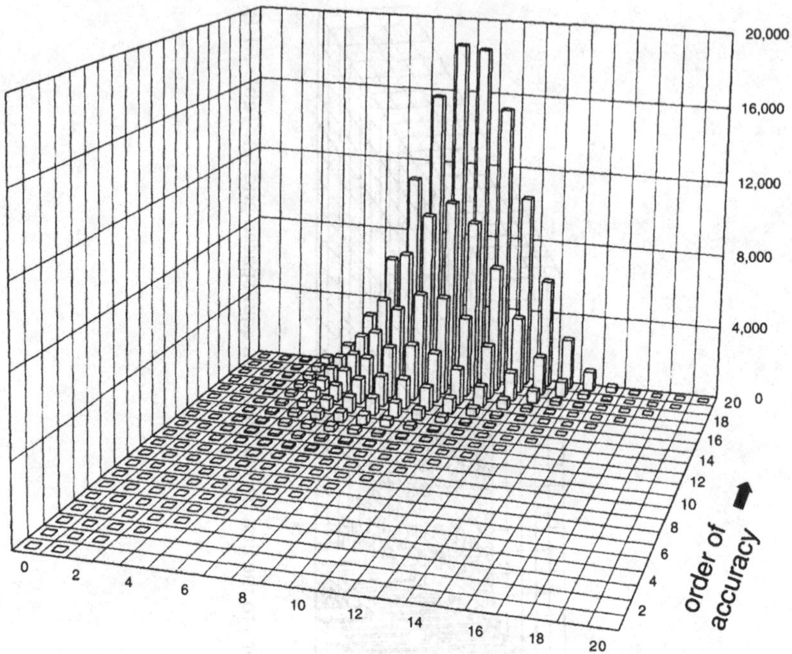

Figure 3.2-2. Magnitude of weights for one-sided approximations to the first derivative on an equi-spaced grid (cf. Table 3.1-2).

$$c^1_{\infty,j} = \begin{cases} \dfrac{(-1)^{j+1}}{j} & \text{if } j = \pm 1, \pm 2, \ldots, \\ 0 & \text{if } j = 0. \end{cases} \tag{3.2-2}$$

Beyond the second derivative, for which $c^2_{p,j} = 2c^1_{p,j}/j$ ($j = \pm 1, \pm 2, \ldots, \pm p/2$) and $c^2_{p,0} = -2\sum_{i=1}^{p/2} 1/i^2$, closed-form expressions for weights become quite complicated. However, this does not affect the ease with which they can be calculated (using the algorithm in Section 3.1) or the existence of simple limits. For the second derivative, the limit becomes

$$c^2_{\infty,j} = \begin{cases} \dfrac{2(-1)^{j+1}}{j^2} & \text{if } j = \pm 1, \pm 2, \ldots, \\ -\dfrac{\pi^2}{3} & \text{if } j = 0. \end{cases}$$

For higher derivatives, the decay rates alternate between $O(1/j)$ and $O(1/j^2)$ for odd and even derivatives, respectively. Formulas for $c^m_{\infty,j}$, $m = 1, 2, \ldots$, are given in Fornberg (1990a).

The situation is very different for one-sided approximations. The closed-form expression for the first derivative is

$$c_{p,j}^{1} = \begin{cases} \dfrac{(-1)^{j+1}}{j}\dbinom{p}{j} & \text{if } j = 1, 2, \ldots, p, \\ -\displaystyle\sum_{i=1}^{p} \frac{1}{i} & \text{if } j = 0. \end{cases} \tag{3.2-3}$$

The magnitudes of the weights form (nearly) a Gaussian distribution, which becomes increasingly peaked at the center of the stencil while growing in height exponentially with p $(\sim \pi^{-1/2} p^{-3/2} 2^{p+3/2})$.

Partly one-sided approximations initially grow more slowly, but will ultimately also diverge exponentially. In the case of the first derivative (just described), the asymptotic rate is multiplied by a factor of $s!/p^s$ if the derivative is evaluated s steps in from the boundary.

One way to derive the closed-form expressions (3.2-1) and (3.2-3) is to write down the general interpolating Lagrange polynomial (with unknown function values at the equi-spaced grid points) and then evaluate its derivative analytically at the point at which we want the formula to be accurate.

3.3. Generalized node distributions

The previous discussion about the size of weights in centered versus one-sided FD formulas suggests that high-order approximations near boundaries will have very large errors. The easiest way to (partly) offset these errors is to concentrate the nodes toward the ends of the interval. The key question becomes: How much grid clustering is called for? Some heuristic arguments point toward *quadratic* clustering - minimum node spacing decreasing like $O(1/N^2)$.

All classical orthogonal polynomials feature quadratic node clustering at the ends. Changing α and β in the weight function $(1-x)^\alpha(1+x)^\beta$ for Jacobi polynomials will still leave the nodes quadratically clustered (this follows e.g. from their differential equation; see Table A-1). Figure 3.3-1 compares the positions of the extrema for Legendre and Chebyshev polynomials of order 20 (corresponding to $\alpha = \beta = 0$ and $\alpha = \beta = -\frac{1}{2}$); there is hardly any noticeable difference.

In the Chebyshev case, the individual terms in Lagrange's interpolation formula do not feature any noticeable spurious oscillations. Given function values $f(x_k)$, $k = 0, 1, \ldots, N$, the unique interpolation polynomial of degree N can be written

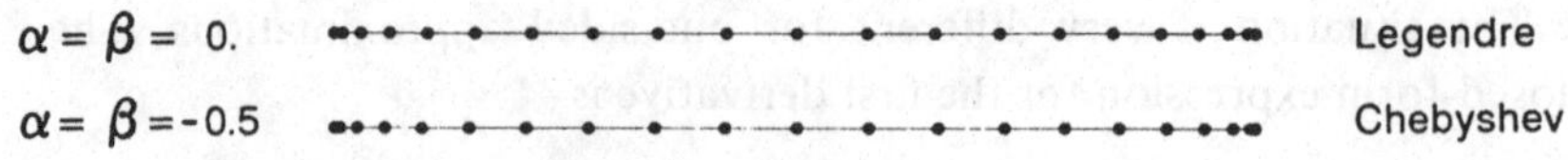

Figure 3.3-1. Difference between location of extrema for Legendre and Chebyshev polynomials ($N = 20$).

$$p_N(x) = \sum_{k=0}^{N} f(x_k)F_k(x), \tag{3.3-1}$$

where

$$F_k(x) = \prod_{\substack{j=0\\j\neq k}}^{N}(x-x_j)\bigg/\prod_{\substack{j=0\\j\neq k}}^{N}(x_k-x_j), \quad k=0,1,\dots,N \tag{3.3-2}$$

are the Nth-degree polynomials satisfying

$$F_k(x_j) = \begin{cases} 1 & \text{if } j=k, \\ 0 & \text{if } j\neq k. \end{cases} \tag{3.3-3}$$

The kth term in (3.3-1) ensures that $p_N(x)$ takes the correct value at the node point x_k without affecting the values at other nodes. The sum will therefore take the desired values at all the nodes.

We can expect the interpolating polynomial $p_N(x)$ to behave nicely *between the nodes* only if the polynomials $F_k(x)$ do so. Figure 3.3-2 shows the difference in this respect between the equi-spaced and the Chebyshev cases. The quadratic grid clustering in the latter case seems to be just strong enough to suppress spurious oscillations near the edges.

Figure 3.3-3 shows $\sum_{k=0}^{N}|F_k(x)|$ (and $\sum_{k=0}^{N} F_k(x) \equiv 1$) for the same cases shown in Figure 3.3-2. The equi-spaced case is again seen to allow far too much spurious oscillation at the ends. The Chebyshev case is much closer to optimality, especially with regard to errors in the first derivative (the nodes are slightly tighter than necessary at the ends if we want only function values).

The Lebesgue constants Λ_N (in the case of $N = 10$) can be read off directly from Figures 3.3-3(a) and (b) because

$$\Lambda_N = \max_{x\in[-1,1]} \sum_{k=0}^{N}|F_k(x)|.$$

Estimates of Lebesgue constants for general N are discussed in Appendix D.

The interpolation error becomes particularly small when the nodes cluster as in the Chebyshev case. A heuristic argument for this goes as follows:

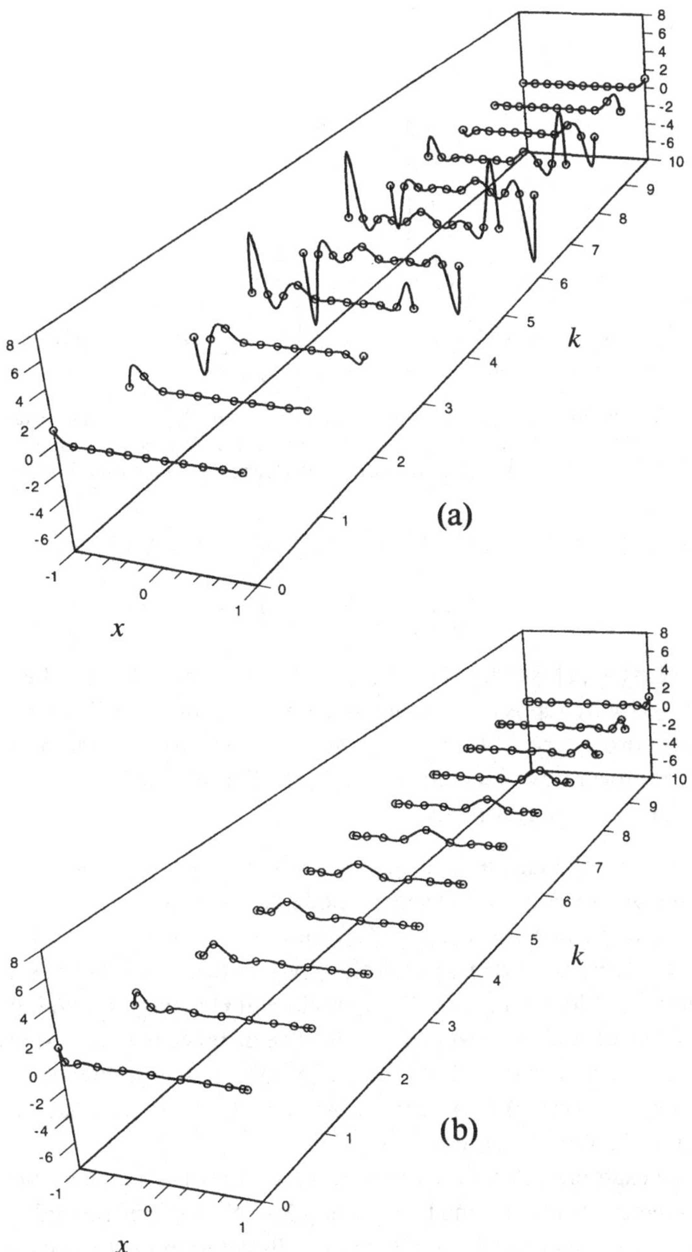

Figure 3.3-2. Basis functions $F_k(x)$ in Lagrange's interpolation formula for two different node distributions in the case of $N = 10$. (**a**) Equi-distributed nodes: $x_i = -1 + 2i/N$, $i = 0, 1, \ldots, N$. (**b**) Chebyshev-distributed nodes: $x_i = -\cos(\pi i/N)$, $i = 0, 1, \ldots, N$.

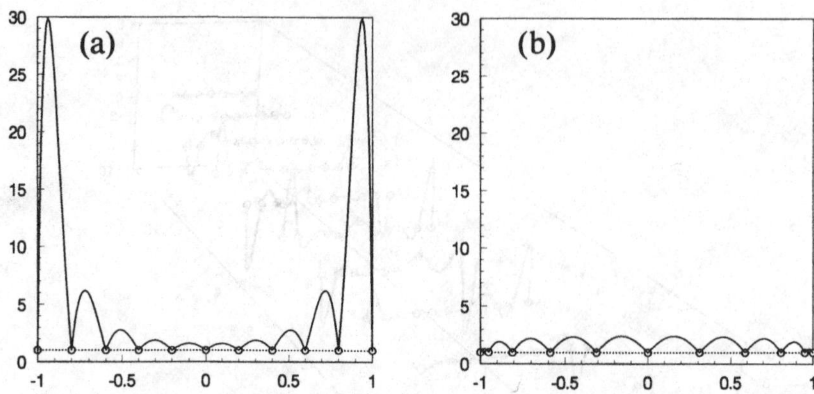

Figure 3.3-3. The functions $\sum_{k=0}^{N}|F_k(x)|$ (solid curves) and $\sum_{k=0}^{N} F_k(x) \equiv 1$ (dashed straight lines) in the case of $N = 10$ (cf. Figure 3.3-2 showing the individual functions $F_k(x)$). **(a)** Equi-distributed nodes. **(b)** Chebyshev-distributed nodes.

The remainder term $R_N(x) = f(x) - p_N(x)$ can be shown to satisfy

$$R_N(x) = \frac{1}{(N+1)!} f^{(N+1)}(\xi) \prod_{j=0}^{N} (x - x_j) \tag{3.3-4}$$

for some $\xi \in [-1, 1]$ (see e.g. Davis 1975). The only part that can be controlled directly by repositioning the nodes x_j is the product. Noting that the highest-order term is $1 \cdot x^{N+1}$, the question becomes: Which polynomial of that form stays smallest over $[-1, 1]$? This is a well-known property of Chebyshev polynomials.

A much more systematic approach to assess how much node clustering is called for starts by considering general node density functions $\mu(x)$. These describe how the *density* of node distributions varies over $[-1, 1]$. It is natural to normalize a density function by $\int_{-1}^{1} \mu(x)\, dx = 1$ because each small interval of length dx will then contain $N\mu(x)\, dx$ nodes, making the total number of nodes $\int_{-1}^{1} N\mu(x)\, dx = N$. The distance between adjacent nodes becomes approximately $1/N\mu(x)$. By considering general $\mu(x)$, we will (in the next section) be able to verify that Chebyshev-type clustering is indeed optimal in the limit of $N \to \infty$.

In some examples, we will select node distributions from the particular one-parameter family outlined in Table 3.3-1. Notice that this table incorporates the equi-spaced and Chebyshev distributions as special cases.

The node density function $\mu_{1/2}(x) = 1/\pi\sqrt{1-x^2}$ is common to all Jacobi polynomials – the differences between them are (in this respect) vanishing as $N \to \infty$. However, the node locations defined by the integral formula in Table 3.3-1 produce only the Chebyshev case ($\alpha = \beta = -\frac{1}{2}$).

Table 3.3-1. *The one-parameter family $\mu_\gamma(x)$ of node density functions*

Density function	Node locations $x_j,\ j=0,1,\dots,N$	Comments
$\mu_\gamma(x)=\dfrac{c_\gamma}{(1-x^2)^\gamma}$	$\dfrac{j}{N}=\int_{-1}^{x_j}\mu_\gamma(x)\,dx$	$\gamma<1;\ c_\gamma=\dfrac{\Gamma(\frac{3}{2}-\gamma)}{\pi^{1/2}\Gamma(1-\gamma)}$
$\mu_0(x)\equiv\dfrac{1}{2}$	$x_j=-1+\dfrac{2j}{N}$	$\gamma=0$; equi-spaced
$\mu_{1/2}(x)=\dfrac{1}{\pi\sqrt{1-x^2}}$	$x_j=-\cos\left(\dfrac{\pi j}{N}\right)$	$\gamma=\dfrac{1}{2}$; Chebyshev

Figure 3.3-4 shows how the nodes move as a function of γ. It also illustrates how relatively small the effect is of changing α and β for Jacobi polynomials.

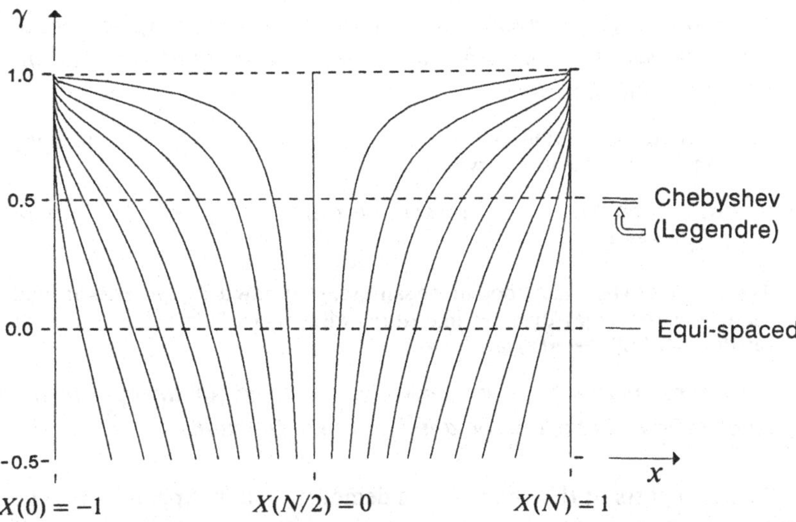

Figure 3.3-4. Distribution of nodes corresponding to density function $\mu_\gamma(x)$, shown for $\gamma\in[-0.5,1]$ ($N=20$). The Legendre distribution (of extrema) is not obtained exactly for any γ, but $\gamma\approx 0.4785$ gives the closest (least squares) fit in this case of $N=20$. This difference is illustrated at the right edge of the figure (it vanishes as $N\to\infty$).

3.4. Rates of convergence and divergence for polynomial interpolation of analytic functions

In a complex ($z = x+iy$) plane, a Taylor series is well known to converge in the *largest circle* around the expansion point that is free of singularities. This result generalizes in a straightforward manner to interpolating polynomials, where the nodes are distributed over an interval rather than all lumped at one point.

Although many books on interpolation theory discuss the convergence of polynomial interpolation, it is surprisingly difficult to locate a simple formulation of the following theorem. General references in this area include Turetskii (1968), Walsh (1960), Krylov (1962), Markushevich (1967), Davis (1975), and Gaier (1987); see also Weideman and Trefethen (1988).

Theorem. *Given a node density function* $\mu(x)$ *(on* $[-1,1]$*), we form the potential function*

$$\phi(z) = -\int_{-1}^{1} \mu(x) \ln|z-x|\, dx \ \ (+\mathit{constant}). \qquad (3.4\text{-}1)$$

Then:

(1) *The polynomial* $p_N(z)$ *(interpolating an analytic function* $f(z)$ *at* N *nodes on* $[-1,1]$*) converges to* $f(z)$ *inside the* largest equi-potential curve *for* $\phi(z)$ *that does not enclose any singularity of* $f(z)$*, and diverges outside that curve.*

Let z_0 denote the limiting singular point of $f(z)$ (or, for now, any point along this largest equi-potential curve).

(2) *The rate of convergence/divergence is exponential, like* $\alpha(z)^N$ *where* $\alpha(z) = e^{\phi(z_0)-\phi(z)}$.

This result is to be understood in the same sense as how a Taylor series around the origin converges/diverges like $\alpha(z)^N$ with $\alpha(z) = |z/z_0|$; i.e., the error satisfies $|R_N(z)|^{1/N} \to \alpha(z)$ as $N \to \infty$.

(3) *PS approximations to any derivative converge (or diverge) in the same fashion as the interpolant does to the function.*

The first two parts of this theorem are demonstrated in Appendix E. The third part is then readily shown by noting that $p_N'(z)$ interpolates $f'(z)$ at N points on $[-1,1]$ separating the $N+1$ points at which $p_N(z)$ interpolated $f(z)$.

Example. Determine the convergence rates at different x positions for equi-spaced interpolation of $f(x) = 1/(1+16x^2)$, $x \in [-1,1]$.

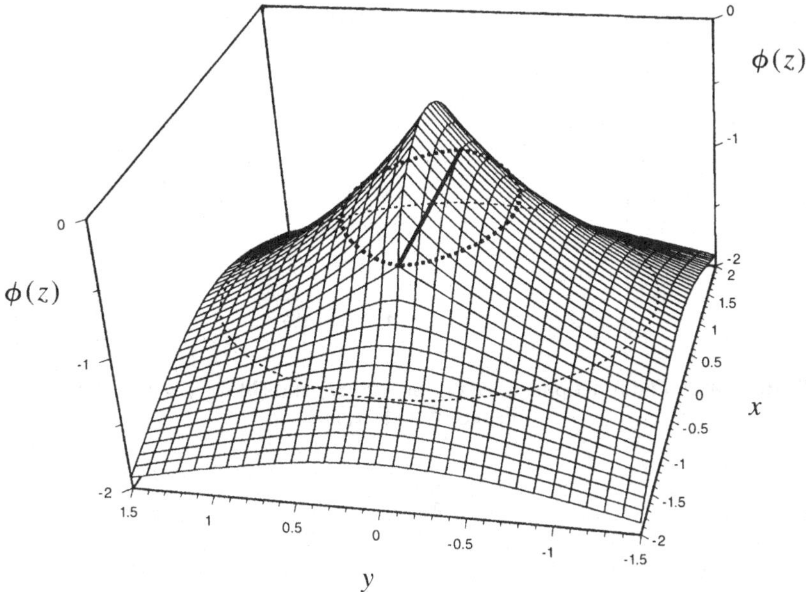

Figure 3.4-1. Potential function $\phi(z)$ for equi-spaced interpolation over $[-1,1]$. The value of $\phi(0)$ is set to zero. The contour lines mark the levels $-\ln 2$ and $-2\ln 2$ (i.e., separations corresponding to factors of 2 in the exponential convergence rate $\alpha(z)$).

In the case of equi-spaced interpolation, $\mu(x) \equiv \frac{1}{2}$. The integral (3.4-1) can be evaluated in closed form as follows:

$$\phi(z) = -\tfrac{1}{2}\,\mathrm{Re}[(1-z)\ln(1-z)-(-1-z)\ln(-1-z)]+C. \quad (3.4\text{-}2)$$

This implies $\phi(0)-\phi(\pm 1)=\ln 2$; i.e., the relation $\alpha(\pm 1)=2\alpha(0)$ always holds for equi-spaced interpolation on $[-1,1]$.

Figure 3.4-1 shows this potential surface (with $C=0$). The heavy (dashed) contour line surrounds the smallest equi-potential domain that includes $[-1,1]$ (on the imaginary axis, it extends just past $\pm 0.5255i$). The function $f(z)$ must be analytic everywhere within this domain for convergence to occur on $[-1,1]$. Any singularity within it restricts convergence to a still smaller equi-potential region, leading to the *Runge phenomenon*: divergence of the interpolant near the ends of the interval.

The function we are interpolating in this example, $f(x)=1/(1+16x^2)$, has only two singularities in the complex plane. They are located at $z=\pm 0.25i$. For $x\in[-1,1]$, from (3.4-2) we have

$$\phi(x)-\phi(\pm 0.25i) = -\tfrac{1}{2}(1-x)\ln(1-x)-\tfrac{1}{2}(1+x)\ln(1+x) + \tfrac{1}{2}\ln\tfrac{17}{16}+\tfrac{1}{4}\arctan(4),$$

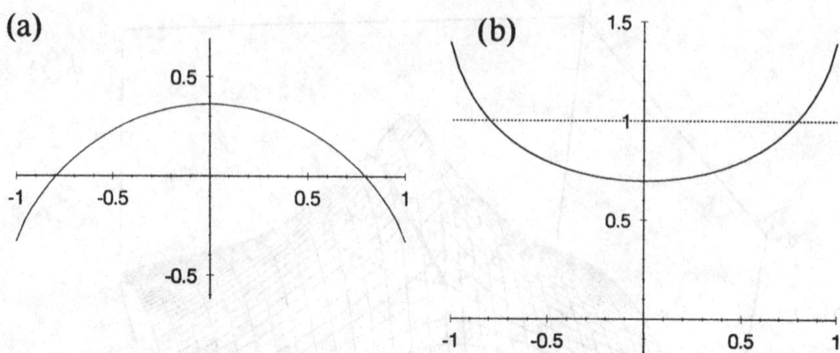

Figure 3.4-2. The functions **(a)** $\phi(x) - \phi(\pm 0.25i)$ and **(b)** $\alpha(x) = e^{\phi(\pm 0.25i) - \phi(x)}$ displayed on $[-1, 1]$.

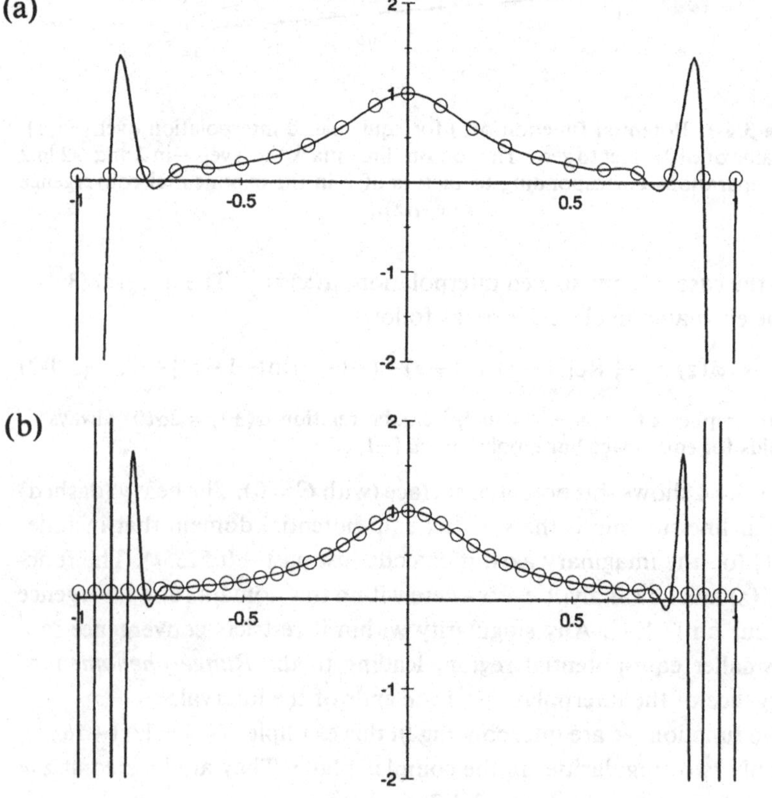

Figure 3.4-3. Results of equi-spaced interpolation on $[-1, 1]$ in the case of **(a)** $N = 20$ and **(b)** $N = 40$.

shown in Figure 3.4-2(a). Figure 3.4-2(b) shows the corresponding convergence rate $\alpha(x)=e^{\phi(\pm 0.25i)-\phi(x)}$. In particular, $\alpha(0)\approx 0.6964$ and $\alpha(\pm 1)=2\alpha(0)\approx 1.3929$. The crossovers between convergence and divergence occur at $x\approx \pm 0.7942$. Figures 3.4-3(a) and (b) show the results of equi-spaced interpolation using $N=20$ and $N=40$, in complete agreement with the predicted rates $\alpha(x)$, and with the (fixed) crossover locations.

Hermite interpolation (requiring not only function values but also first-derivative values to match at the gridpoints) offers no help against the Runge phenomenon. Such interpolation converges and diverges in precisely the same places as does Lagrange interpolation.

To return to the question of the optimal $\mu(x)$, we compare in Figure 3.4-4 the potential surfaces $\phi(z)$ corresponding to different density functions $\mu_\gamma(x)$ as described in Section 3.3. The case $\gamma=0$ is the same as shown in Figure 3.4-1 (but with different contours marked).

For Jacobi polynomials (sharing the same node density function $\mu_{0.5}(x)=1/\pi\sqrt{1-x^2}$), the integral in (3.4-1) can again be evaluated in closed form:

$$\phi(z)=-\ln|z+\sqrt{z^2-1}|+C. \tag{3.4-3}$$

Like (3.4-2), this expression is correct for all complex values of z when selecting the conventional branches for the logarithm and (here) the square root functions.

The potential surface in this case forms a perfectly flat ridge along $[-1,1]$. This is clearly optimal - it represents the only possibility of convergence on $[-1,1]$ that does not require analyticity anywhere outside the interval $[-1,1]$. Figure 3.4-4(c) shows what happens when the nodes cluster denser still toward the ends ($\gamma=0.7$). To converge near $z=0$, $f(z)$ needs to be analytic in areas surrounding $z=\pm 1$.

Figure 3.2-2 showed that the difficulty with approximations at boundaries increased in severity very rapidly with N. The results in Sections 3.3 and 3.4, pointing to Chebyshev-type node clustering as optimal, have been concerned only with the limit of $N\to\infty$. This leaves open the possibility that a less severe clustering would suffice for finite N. Thanks to the FD approach to PS methods, this question can be addressed (see the additional comments in Sections 5.1 and 5.6).

3.5. Example of a differentiation matrix

We consider the same situation as in Section 2.3, namely, the first derivative approximated on a periodic, equi-spaced grid. Instead of using trigonometric interpolation, we employ the limiting FD formula (3.2-2). This

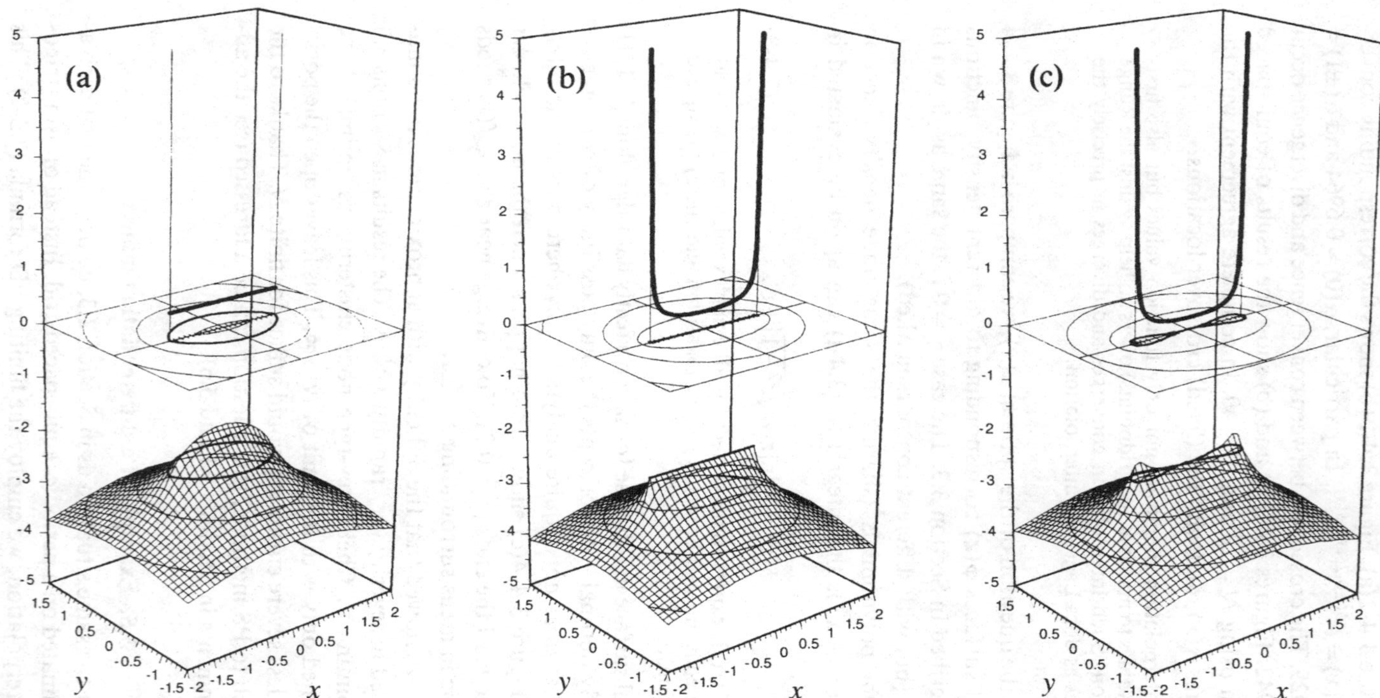

Figure 3.4-4. Node density functions $\mu_\gamma(x)$ and their corresponding logarithmic potentials $\phi_\gamma(x, y)$. The heavy contour lines mark the regions that must be free from singularities for convergence to occur (over $[-1, 1]$) as the number of nodes $N \to \infty$. Lower contour lines (vertically separated by 0.5) correspond to geometric convergence rates α^N, with $\alpha = e^{-0.5} \approx 0.607$, $e^{-1.0} \approx 0.368$, $e^{-1.5} \approx 0.223, \dots$. **(a)** Equi-spaced, $\gamma = 0.0$. **(b)** Jacobi, $\gamma = 0.5$. **(c)** $\gamma = 0.7$.

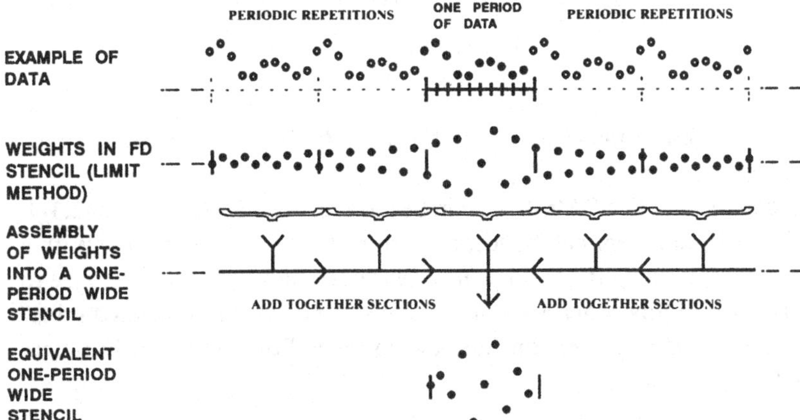

Figure 3.5-1. Application of limiting FD method to periodic data.

formula has infinite width, but its application is possible because periodic data can be thought of as repeating indefinitely. Again assuming $N = 2m+1$ and adjusting the weights for a mesh spacing of $h = 2/N$ (rather than $h = 1$ as in (3.2-2)), we obtain

$$c^1_{\infty,j} = \begin{cases} \dfrac{N}{2}\dfrac{(-1)^{j+1}}{j} & \text{if } j \neq 0, \\ 0 & \text{if } j = 0. \end{cases}$$

Figure 3.5-1 illustrates how period-wide sections of the stencil can be added together to create an equivalent stencil covering only one period of the data. The weights then become

$$\begin{aligned} d^1_{\infty,j} &= \frac{N}{2}(-1)^{j+1} \sum_{k=-\infty}^{\infty} \frac{(-1)^k}{j+Nk} \\ &= \frac{1}{2}(-1)^{j+1} \sum_{k=-\infty}^{\infty} \frac{(-1)^k}{k+(j/N)} \quad \text{if } j = \pm 1, \pm 2, \dots, \pm m, \\ & \qquad\qquad\qquad\qquad\qquad\qquad (\pm(m+1), \dots, \pm(N-1)) \\ d^1_{\infty,j} &= 0 \qquad\qquad\qquad\qquad\qquad \text{if } j = 0. \end{aligned}$$

The DM is a cyclic matrix. Its (i,j)th element $D_{i,j}$ is $d^1_{\infty,i-j}$. Noting the identity $\sum_{k=-\infty}^{\infty}[(-1)^k/(k+x)] = \pi/(\sin \pi x)$, we have

$$D_{i,j} = \begin{cases} \dfrac{\pi(-1)^{i-j}}{2\sin(\pi(i-j)/N)} & \text{if } i \neq j, \\ 0 & \text{if } i = j. \end{cases} \quad (3.5\text{-}1)$$

3.6. Equivalence of PS methods and limits of FD methods

Periodic case. The DMs derived in Sections 2.3 and 3.5 (equations (2.3-4) and (3.5-1)) are identical; hence, the two methods are equivalent. With a modest additional effort, this equivalence can be shown to generalize to derivatives of any order, to even numbers of points, to "staggered grids" (discussed further in Section 5.3), and so forth. For details, see Fornberg (1990a).

Nonperiodic case. We assume that the data cannot be extended past the boundaries. The order of accuracy for the approximations correspond to the number of gridpoints (rather than being formally infinite). The PS method now turns out to be equivalent to using the FD approximations, whose stencils extend over all the gridpoints. This can be seen as follows.

PS approach: Consider data given at $N+1$ points on $[-1,1]$ (distributed e.g. according to the zeros or extrema of some orthogonal polynomial, as is customary in PS collocation; however, their distribution is irrelevant for our present argument). By means of expansion in these polynomials, the PS method provides the exact derivative of the interpolation polynomial going through the data at these points.

FD approach: With no data extensions available, we can at best consider FD stencils that are as wide as the grid. To approximate derivatives at the gridpoints, the FD weights must be calculated separately for each point. Every one of these FD stencils will give the exact result for any Nth-degree polynomial.

For any given data and distribution of nodes, the interpolating polynomial of minimal degree is unique. Since both approaches give the exact results for this polynomial, these results will always be the same. Hence, the approaches are equivalent.

High-order polynomial interpolation has a bad reputation because of the Runge phenomenon (see Sections 3.3 and 3.4) and the following theorem: *For any node density function, one can design a continuous function such that the max-norm error tends to infinity as the number of nodes is increased.*

This bad reputation is not always justified. The highly successful nonperiodic PS method amounts to taking exact derivatives of very high-order interpolation polynomials. Interpolation using the Chebyshev (or similar) nodes features no Runge phenomenon. The low Lebesgue constant $O(\log N)$ ensures that, even for the most artificially constructed functions (e.g., those that are not continuous at any point), high-order interpolation still leads to a polynomial that is not too far off from the optimal one (and that is able to approximate most functions very closely).

4

Key properties of PS approximations

In the previous chapters, we have repeatedly referred to the exponential convergence rate of spectral methods for analytic functions. This is discussed in more detail in Section 4.1. When functions are not smooth, PS theory is much less clear. An approximation can appear very good in one norm and, at the same time, very bad in another. As illustrated in Section 4.2, PS performance can also be very impressive in many cases that are "theoretically questionable" - this is exploited in most major PS applications. Sections 4.3–4.5 describe differentiation matrices in more detail, their influence on time stepping procedures, and linear stability conditions. A fundamentally different kind of instability, specific to nonlinear equations, is discussed in Section 4.6. Very particular distributions of the nodes yield spectacular accuracies for Gaussian quadrature formulas. PS methods are often based on such formulas, presumably with the hope of obtaining a correspondingly enhanced accuracy when approximating differential equations. In the examples of Section 4.7, we see little evidence for this.

Smoothness of a function is a rather vague concept. Increasingly severe requirements include:

- a finite number of continuous derivatives;
- infinitely many derivatives; and
- analyticity - allowing continuation as a differentiable complex function away from the real axis.

In the limit of N (number of nodes or gridpoints) tending to infinity, these cases give different asymptotic convergence rates for PS methods. In the first case, the rate becomes polynomial with the power corresponding to the number of derivatives that are available. In this "practical guide" we focus on the two extreme cases: analytic functions in Section 4.1, and discontinuous functions in Section 4.2. For intermediate requirements, intermediate results can be obtained. The reviews mentioned in the introduction to Chapter 2 discuss these cases in some detail. However, it should be noted that for finite N – which is the case in all calculations, since data are available at only a finite number of gridpoints – even the difference between analytic and discontinuous functions can become somewhat unclear. Infinitely many analytic functions will perfectly fit any finite discrete data set. On the other hand, data from a genuinely analytic function can "look" very rough on the coarse grids that often suffice when using PS methods.

4.1. Convergence of PS methods for smooth functions

Nonperiodic case

Polynomial interpolation of smooth functions based on the Chebyshev nodes (as well as expansions in Chebyshev polynomials) are well known to provide approximations with nearly uniform accuracy over $[-1,1]$, whereas interpolation based on equi-spaced points can diverge near the ends (the Runge phenomenon). The potential functions described in Section 3.4 provided a general tool for addressing such issues as when, and with what rates, the convergence will occur.

The relationship between the potential contours and the rates of convergence becomes particularly simple in the Chebyshev case. Some algebraic manipulation of (3.4-3) reveals that the contour corresponding to a convergence rate α^N ($\alpha \in (0,1)$, $x \in [-1,1]$) is the ellipse

$$\frac{x^2}{((\alpha+1/\alpha)/2)^2}+\frac{y^2}{((\alpha-1/\alpha)/2)^2}=1, \tag{4.1-1}$$

with foci at ± 1. Conversely, given the location of the nearest singularity, this equation gives the convergence factor α. Equation (4.1-1) can also be derived more directly from Lagrange's interpolation formula and by noting that $T_n(x)=\frac{1}{2}(z^n+1/z^n)$ where $\frac{1}{2}(z+1/z)=x$. If here we consider x and z as complex variables then the ellipses (4.1-1) in the x plane correspond to circles in the z plane, centered at the origin and with radii $1/\alpha$.

To illustrate how the convergence of polynomial interpolants depends on the properties of a smooth function $f(x)$ (f and x are now real), let us consider the two-parameter class of functions

$$f_{\xi,\eta}(x)=\frac{1}{1+((x-\xi)/\eta)^2}.$$

Graphically, these functions have a "hump" of unit height, centered at $x=\xi$, with widths (and radius of curvature at the tips) proportional to η.

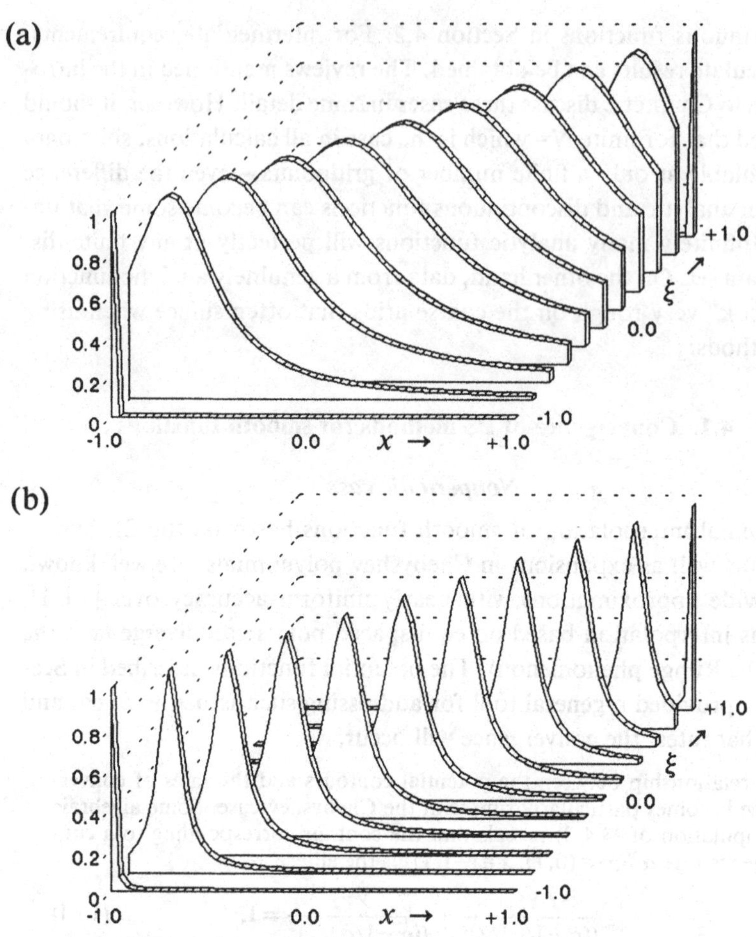

Figure 4.1-1. Functions $f_{\xi,\eta}(x) = 1/(1+[(x-\xi)/\eta]^2)$ with minimal η (i.e. maximal curvature at the tip). **(a)** Equi-spaced nonperiodic PS: borderline convergence/divergence as $N \to \infty$. **(b)** Chebyshev PS: convergence rate α^N with $\alpha = 0.9$.

The equi-spaced PS method will be just borderline converging/diverging when the closest singularity of $f_{\xi,\eta}(x)$ (which has only two singularities, located at $x = \xi \pm i\eta$) falls on the equi-potential curve passing through $x = \pm 1$ (drawn bold in Figure 3.4-1). Figure 4.1-1(a) shows these most peaked functions $f_{\xi,\eta}(x)$ for different values of $\xi \in [-1,1]$. The equi-spaced PS method is clearly much better able to resolve high curvatures near the ends of $[-1,1]$ than in the interior. However, it is also clear that (in this form) the equi-spaced nonperiodic PS method is quite useless – no matter how many points are employed, it will not converge over $[-1,1]$ for functions any more peaked than those shown in Figure 4.1-1(a).

This issue is discussed also by Solomonoff and Turkel (1989). They quote a different formulation of (3.4-2) but select an inappropriate branch, leading to some flawed results.

Interpolation at the Chebyshev nodes will work for any ξ and any $\eta > 0$. However, if we require a "reasonable" convergence rate, say $\alpha = 0.9$ (i.e., a factor of approximately 10^{-6} for every 130 node points), then the situation is again somewhat similar; see Figure 4.1-1(b). Once more, the highest resolution is obtained near the boundaries. This feature has been exploited frequently - for example, to resolve boundary layers in fluid mechanics. Note, however, that this is *not* a consequence of the grid being finer near the boundary (this effect was no less prominent in the case of equi-spaced grids).

Periodic case

For the periodic PS method (on $[-1, 1]$), the formula corresponding to (4.1-1) becomes

$$y = \pm \frac{2}{\pi} \ln \alpha. \tag{4.1-2}$$

This is related to the fact that a Fourier series converges in the widest horizontal strip around the x axis that is free of singularities.

General discussion

Figure 4.1-2 compares the curves given by equations (4.1-1) and (4.1-2) for $\alpha = 0.5$ and $\alpha = 0.9$.

In the limit of $\alpha \to 1$, the crossover points tend to $x = \pm(1 - 4/\pi^2) \approx \pm 0.595$. For functions with very fine detail, the periodic and the nonperiodic PS methods are therefore rather comparable. With the same number of points over $[-1, 1]$, the former achieves higher accuracy within the central section $[-0.595, 0.595]$ and the latter in $[-1, -0.595]$ and $[0.595, 1]$. The break-even points in local grid densities occur at $x \approx \pm 0.561$.

In the "opposite" limit of $\alpha \to 0$ (i.e., cases of very fast geometric convergence), the ellipses in Figure 4.1-2 expand in size like $O(1/\alpha)$ whereas the parallel strips widen only like $O(\log \alpha)$. For very smooth functions, the periodic PS method converges much faster than its nonperiodic counterpart.

In every instance of PS methods applied to functions analytic in some neighborhood of $[-1, 1]$, the convergence takes the form $O(\alpha^N)$ (discussed so far only for interpolation, but clearly true as well - with the same α - for approximations to any derivative). This rate distinguishes spectral methods from FD and FE methods (where the rate for a pth-order method

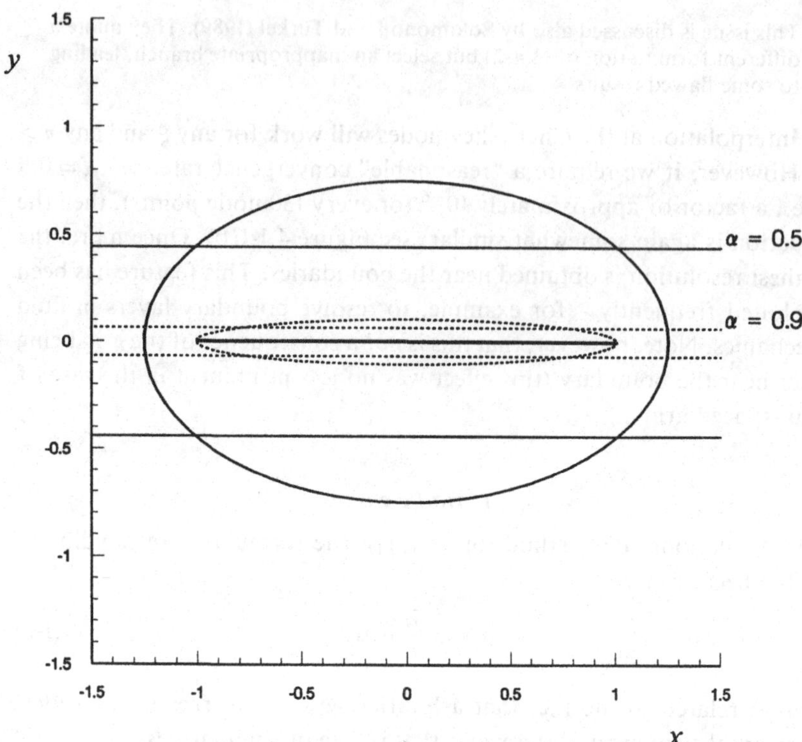

Figure 4.1-2. Comparisons of domains in complex x plane that need to be free of singularities to obtain convergence rates α^N, $\alpha = 0.5$ and 0.9. Chebyshev method: ellipses with foci ±1 (eq. 4.1-1). Periodic PS method: horizontal strips around the x axis (eq. 4.1-2).

would be $O(1/N^p)$, reflecting polynomial rather than exponential convergence). Whether or not there exists a classical family of orthogonal polynomials associated with the PS method is quite irrelevant.

The relative resolution ability of different methods is sometimes expressed as the number of points needed per wavelength. For a Fourier expansion, this number is 2 (Kreiss and Oliger 1972). For nonperiodic PS methods, it is π in the Chebyshev case (Gottlieb and Orszag 1977) and 6 in the equi-spaced case (Weideman and Trefethen 1988; via a simpler argument in Trefethen and Weideman 1991).

The result of 6 points per wavelength for nonperiodic equi-spaced grids is of theoretical interest only. Owing to the extremely high Lebesgue constant in this case, rounding errors will be magnified and thus will likely ruin any attempts to reach high accuracies.

Here we can indicate one starting point for an analysis comparing FD and PS methods. We assume periodicity, and that $N+1$ gridpoints are equi-spaced $h=2/N$ apart within the period $[-1,1]$. The range of Fourier modes $e^{i\omega x}$ that can be represented on this grid is $-\omega_{\max} \le \omega \le \omega_{\max}$, where $\omega_{\max} = \pi/h$.

Any mode ω outside $[-\omega_{\max}, \omega_{\max}]$ will, on the grid, appear equivalent to a mode within this range - an *aliasing* error (see Section 4.6). Both the number of different modes that are present - and their magnitude - depend on the regularity of the function we approximate.

Suppose we want to approximate d/dx. For a mode $e^{i\omega x}$, the exact answer should be

$$\frac{d}{dx}e^{i\omega x} = i\omega e^{i\omega x}.$$

With FD2 (second-order finite differences) we have

$$D^{(2)}e^{i\omega x} = \frac{e^{i\omega(x+h)} - e^{i\omega(x-h)}}{2h} = i\frac{\sin \omega h}{h}e^{i\omega x} = if(2,\omega,h)e^{i\omega x}. \qquad (4.1\text{-}3)$$

For the centered pth-order FD schemes $D^{(p)}$, $p = 2,4,6,\ldots$, we similarly obtain

$$f(p,\omega,h) = \left\{\frac{\sin \omega h}{h}\sum_{k=0}^{p/2-1}\frac{(k!)^2}{(2k+1)!}2^{2k}\left(\sin\frac{\omega h}{2}\right)^{2k}\right\}. \qquad (4.1\text{-}4)$$

To derive (4.1-4), we write

$$D^{(p)} = D_0\sum_{k=0}^{p/2-1}(-1)^k\alpha_k(h^2D_+D_-)^k,$$

where $D_0 f(x) = [f(x+h)-f(x-h)]/2h$, $D_+f(x) = [f(x+h)-f(x)]/h$, and $D_-f(x) = [f(x)-f(x-h)]/h$. Applying the operator $D^{(p)}$ to $e^{i\omega x}$ gives

$$D^{(p)}e^{i\omega x} = i\omega + O(\omega^{p+1}h^p) = \frac{i}{h}\sin \omega h\sum_{k=0}^{p/2-1}\alpha_k 2^{2k}\left(\sin\frac{\omega h}{2}\right)^{2k}e^{i\omega x},$$

of the same form as (4.1-4). Substituting $\sin(\omega h/2) = \xi$ gives

$$\frac{\arcsin \xi}{\sqrt{1-\xi^2}} = \xi\sum_{k=0}^{p/2-1}\alpha_k 2^{2k}\xi^{2k} + O(\xi^{p+1}).$$

The coefficients α_k follow (recursively) from noting that

$$(1-\xi^2)\frac{\partial}{\partial\xi}\left(\frac{\arcsin\xi}{\sqrt{1-\xi^2}}\right) = 1+\xi\left(\frac{\arcsin\xi}{\sqrt{1-\xi^2}}\right).$$

Figure 4.1-3 compares $f(p,\omega,h)$ to the exact result ω. For $p=2$, only a fraction of the Fourier modes present are treated even nearly correctly.

As $p\to\infty$, convergence is seen to occur like that of a Taylor expansion: the number of correct derivatives at the origin is the same as the order of the

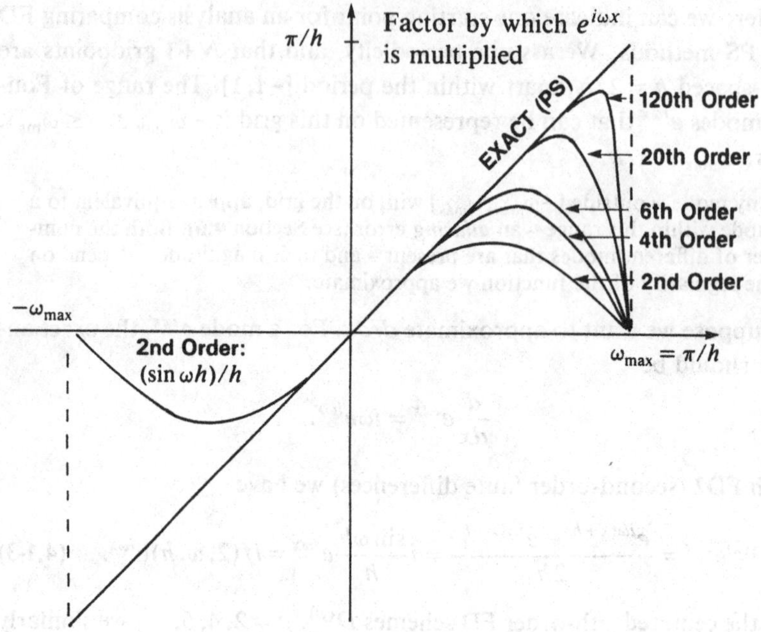

Figure 4.1-3. Multiplicative factors $f(p,\omega,h)$ arising when the pth-order FD approximation for d/dx is applied to $e^{i\omega x}$.

FD scheme. It may make sense to give up some of the (often unnecessarily high) accuracy for low ω (i.e. for long waves) in exchange for keeping the accuracy within some uniform tolerance over a wider range of ω (or over a specific narrow frequency band relevant to a particular application, such as modeling vibrator-generated waves for seismic exploration). Compact FD schemes of this type are described by Holberg (1987), Mittet et al. (1988), and Kindelan, Kamel, and Sguazzero (1990).

Solomonoff (1994) presents still another approach for generating FD schemes that are optimized in application-specific ways (i.e., rather than being designed to be exact for polynomials of as high order as possible). He notes that such schemes can be made less vulnerable to the Runge phenomenon.

Lele (1992) displays figures similar to Figure 4.1-3 also for many implicit (compact) derivative approximations, such as

$$[\tfrac{1}{6}f'(x-h)+\tfrac{2}{3}f'(x)+\tfrac{1}{6}f'(x+h)] = [-\tfrac{1}{2}f(x-h)+\tfrac{1}{2}f(x+h)]/h+O(h^4),$$

and notes that these can offer some advantages over their explicit counterparts – in this case,

$$f'(x) = [\tfrac{1}{12}f(x-2h)-\tfrac{2}{3}f(x-h)+\tfrac{2}{3}f(x+h)-\tfrac{1}{12}f(x+2h)]/h+O(h^4).$$

Lele also surveys the history of compact FD schemes.

In the $p=\infty$ limit (the periodic PS method), the only errors are aliasing errors (discussed further in Section 4.6). Because of variable coefficients and/or nonlinearities, high Fourier modes outside the range of $[-\omega_{\max}, \omega_{\max}]$ are generated and then possibly mistreated. One approach for controlling such errors is to apply weak damping (dissipation). However, as we shall see, anyone who views aliasing solely as a source of errors is missing out on one of the most important (and intriguing) strengths of PS methods.

4.2. Convergence of PS methods for nonsmooth functions

The PS method often performs extremely well even in cases where both solutions and variable coefficients are not smooth. Several of the major PS applications depend on this (turbulence modeling, weather forecasting, seismic modeling, etc.).

Example. Compare FD2, FD4, and PS solutions to the $[-1,1]$-periodic 1-D acoustic wave equation

$$\begin{cases} u_t = v_x, \\ v_t = c^2(x) u_x, \end{cases} \quad \text{where } c(x) = \begin{cases} 1 & \text{if } -1 < x < 0, \\ \frac{1}{2} & \text{if } 0 < x < 1, \end{cases} \tag{4.2-1}$$

with a sharp pulse as the initial condition.

In each of the intervals $[-1,0]$ and $[0,1]$, equation (4.2-1) supports solutions of the following forms, traveling to the right ($\Rightarrow$) and to the left ($\Leftarrow$):

$$\text{in } [-1,0], \quad \begin{cases} u(x,t) = -v(x,t) & \Rightarrow, \\ u(x,t) = +v(x,t) & \Leftarrow; \end{cases}$$

$$\text{in } [0,1], \quad \begin{cases} u(x,t) = -2v(x,t) & \Rightarrow, \\ u(x,t) = +2v(x,t) & \Leftarrow. \end{cases}$$

We choose as the initial condition $u(x,0) = 2v(x,0) = e^{-1600(x-1/4)^2}$. Figure 4.2-1 shows the time evolution for $u(x,t)$; the one for $v(x,t)$ is qualitatively similar. After the pulses have hit the interfaces at $x = 0$ and $x = \pm 1$ numerous times – on each occasion generating two outgoing pulses (transmitted and reflected) – the analytic solution at $t = 7$ consists of just three pulses.

The PS method and the periodic second- and fourth-order FD methods give, at $t = 7$, the results shown in Figure 4.2-2.

We have used grids where 0 and ± 1 fall halfway between gridpoints, thus saving us from deciding on the values of $c(x)$ at these locations. No numerical smoothing was used in any of these cases. The time integration was performed with leapfrog (FD2 in time) with sufficiently small time step that the observed errors are all due to the spatial discretizations. Many other ODE

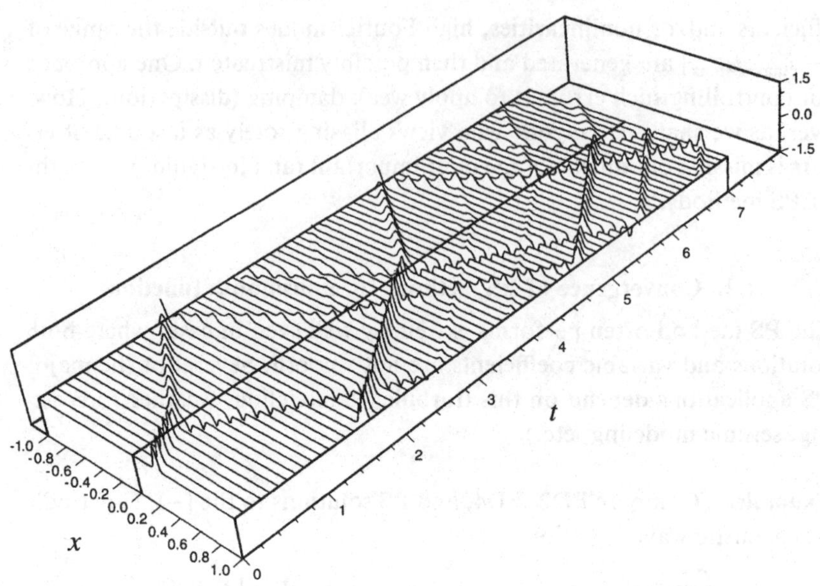

Figure 4.2-1. Time evolution of $u(x,t)$ solving equation (4.2-1).

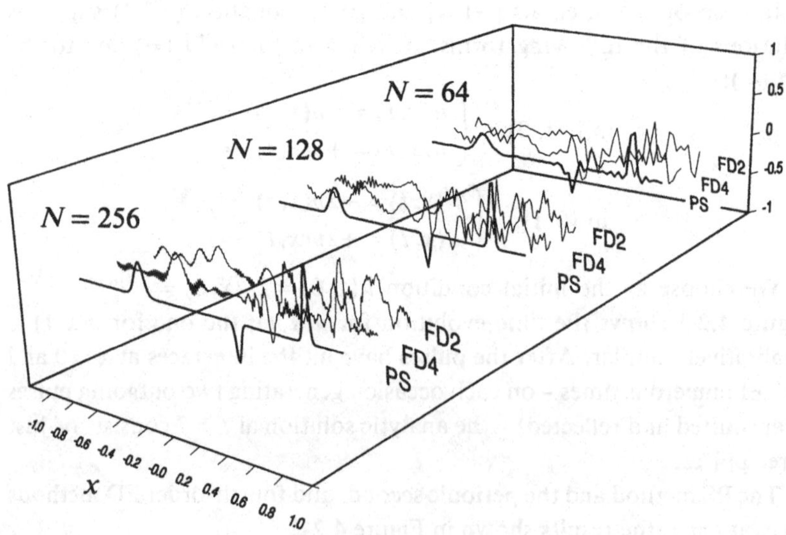

Figure 4.2-2. Comparison of numerical solutions for $u(x,t)$ at time $t = 7$ using different methods and grid densities.

solvers could have been used equally well (such as Runge–Kutta, AB/AM predictor–corrector, etc.; see Section 4.5 and Appendix G).

We can note the following.

- Already with $N = 64$, the PS method has retained considerable accuracy (in spite of the initial pulse being only about two gridpoints wide). For higher values of N, the performance of the PS method is flawless, and far superior to that of the FD methods.
- One might have expected the PS method to develop Gibbs-type oscillations. Instead, it is the FD methods that develop problems with the high modes. The generally used remedy against spurious high-frequency oscillations is to apply some viscous damping – preferably as little as possible, to avoid smearing out the pulses themselves. This example shows that, for the PS method, a very small amount of damping would often be sufficient (although to use none, as in this example, is unnecessarily risky).

Trefethen (1982) observes similar-looking high-frequency oscillations in an equivalent leapfrog–FD4 calculation for the one-way wave equation with a discontinuous medium. He notes that the leapfrog time differencing scheme allows such high-frequency "parasite" solutions to be transported in the opposite direction to the main wave.

As Figure 4.2-3 shows, the difference between the FD2, FD4, and PS methods lies not in the size of the local errors, but in how such errors accumulate or cancel over time. The figure illustrates how step solutions preserve their steepness over time (and similarly, how other nonsmooth solutions preserve their distinctive features) far better when high-order FD or PS methods rather than low-order FD methods are used. The FD2 approximation to the derivative of a step function may look more accurate, but this circumstance is irrelevant to its performance during time integration.

Both the FD2 and the FD4 errors are *dispersive*. In the context of solving $\partial u/\partial t + \partial u/\partial x = 0$, the curves in Figure 4.1-3 can be interpreted as the angular phase speed of different modes. For low-order methods, high modes travel too slowly, thus giving rise to the trailing oscillations seen in Figure 4.2-3.

The estimate $O(1/\sqrt[p+1]{t})$ for the slopes in Figure 4.2-3 can be arrived at as follows: The FDp approximation to $\partial/\partial x$ has $ch^p(\partial^{p+1}/\partial x^{p+1})$ as its leading error term; i.e., we are more closely approximating an equation $\partial u/\partial t + \partial u/\partial x + ch^p(\partial^{p+1}u/\partial x^{p+1}) = 0$ than $\partial u/\partial t + \partial u/\partial x = 0$. By discarding the translation term $\partial u/\partial x$ and choosing as initial conditions the step function

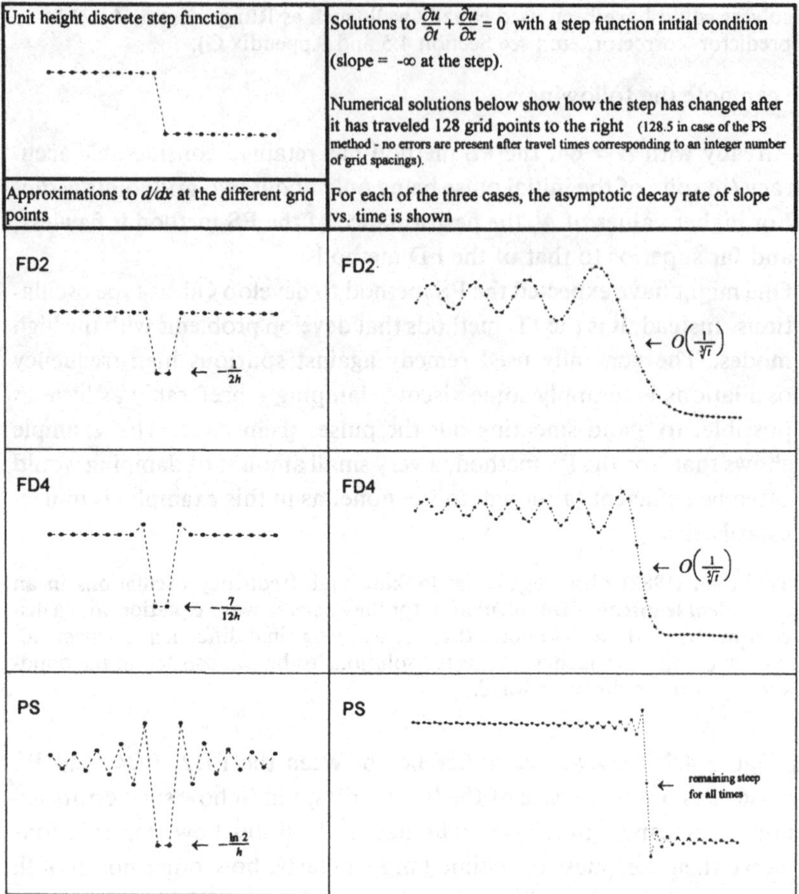

Figure 4.2-3. Errors when approximating the derivative of a step function: local approximation to the derivative near a step versus the long-term evolution of numerical solutions with a step initial condition.

$$u(x,0) = 1 - H(x) = \begin{cases} 1 & \text{if } x < 0, \\ 0 & \text{if } x \geq 0, \end{cases}$$

the analytical solution to the new equation becomes

$$u(x,t) = \frac{1}{2} - \frac{1}{2\pi}\int_{-\infty}^{\infty} \frac{1}{\omega}[(\cos ch^p\omega^{p+1}t)(\sin \omega x) + (\sin ch^p\omega^{p+1}t)(\cos \omega x)]\, d\omega.$$

At $x = 0$, its x derivative is

$$\left.\frac{\partial u}{\partial x}\right|_{x=0} = -\frac{1}{2\pi}\int_{-\infty}^{\infty} (\cos ch^p\omega^{p+1}t)\, d\omega =$$

$$= -\frac{1}{\pi(p+1)}\cos\left(\frac{\pi}{2(p+1)}\right)\Gamma\left(\frac{1}{p+1}\right)\Big/\sqrt[p+1]{ch^p t} = O(1/\sqrt[p+1]{t}).$$

In order for the slope of a step to be reduced by a factor of 2, the time t must be increased by a factor of 8 when using FD2 and by over 2000 when using FD10.

Figure 4.2-2 shows not only that nonsmooth solutions translate very accurately when using high-order FD/PS methods in cases of a constant medium, but also that reflections and transmissions are very accurate when the medium is discontinuous. At present, there appears to be no satisfactory theoretical argument supporting this (important) fact.

In Fornberg (1987, 1988a), many tests similar to the preceding example (with the 1-D acoustic wave equation) were carried out for the 2-D elastic wave equation – with very similar results. See the discussion and illustrations in Section 8.4.

If Gibbs' oscillations do arise in PS calculations, spectral pointwise accuracy can still sometimes be recovered in smooth regions by post-processing. Gresho and Lee's (1981) paper, "Don't suppress the wiggles – they're trying to tell you something", discusses this, and points out that viscous damping during a calculation (even when applied with good intent) can lead to an irretrievable loss of information. Although the convergence for step solutions is bad in most common norms (Majda, McDonough, and Osher 1978), Abarbanel, Gottlieb, and Tadmor (1986) note that this need not be the case for certain "negative Sobolev norms". Spectral accuracy of "moments" provides information that can be used to restore Gibbs-affected solutions.

For many nonlinear equations, the discontinuities that arise are not of a "shock" type but are rather like contact discontinuities, which are quite passively transported in a linear fashion (as with direct simulations of turbulence and, to a lesser extent, with weather forecasting; see Sections 8.1 and 8.3). In such cases, it may suffice to add very little viscosity and rely on the method's ability to handle linear situations.

The problem with more severe nonlinearities is primarily that they can introduce couplings, disrupting the delicate cancellation process upon which the PS method for nonsmooth functions depends. One idea is to add some more viscosity (just enough to be able to exploit the PS method's power in case of smooth solutions, but not so much that the solution itself is too severely affected). Spectrally accurate solutions can sometimes be obtained in this way even for shock problems. This is an old idea for which different versions have been proposed; one is the *spectral viscosity* method of Tadmor (1989). Further discussions on this and similar methods can be found in Cai, Gottlieb, and Shu (1989, 1992), Tadmor (1990, 1993), and Maday, Kaber, and Tadmor (1993).

4.3. Differentiation matrices

For both computational and theoretical purposes, it is often convenient to collect all the weights for the approximations at the gridpoints in a *differentiation matrix* (DM; see Sections 2.3 and 3.5). Finding the derivative of a vector of data becomes a matrix × vector multiplication.

The relative efficiencies of straightforward matrix × vector multiplications ($O(n^2)$ operations) versus FFT-based Chebyshev recursions ($O(n \log n)$ operations) have been compared many times. The estimates for the break-even

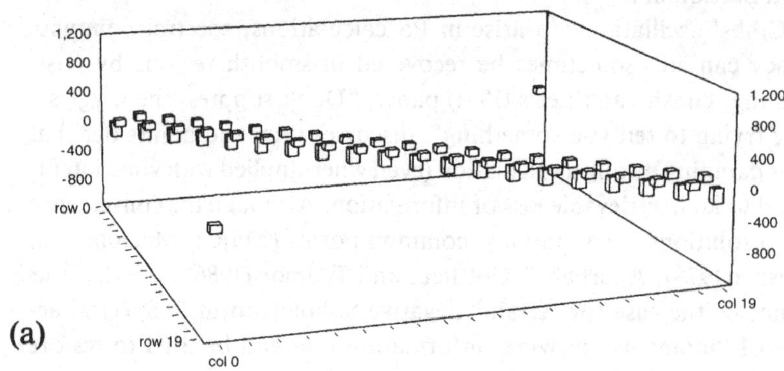

(a)

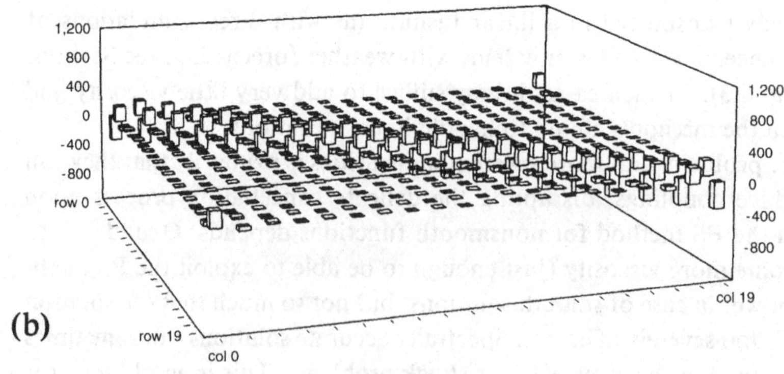

(b)

Figure 4.3-1. Differentiation matrices for approximations to d^2/dx^2, $N = 20$ (in brackets, maximum magnitude of DM element). **(a)** FD2, periodic [200]. **(b)** PS, periodic [331].

point range at least from $n = 16$ (Canuto et al. 1988) to $n \approx 100$ (Taylor, Hirsh, and Nadworny 1984). As Figure F.3-1 shows, the point can be higher still for vector and parallel machines. Furthermore, Solomonoff (1992) shows how a restructuring of the matrix × vector multiplication for DMs can nearly double the speed of this approach.

The *fast multi-pole* method achieves an $O(n \log n)$ operation count for arbitrary node distributions (Boyd 1992). However, the proportionality constant is much higher than for FFTs, so the approach appears not to be competitive in the present context.

Figure 4.3-1 illustrates what the DMs look like for the second derivative in the case of $N = 20$ (i.e., 21 gridpoints if both ends are included, 20 within the period for periodic problems). Part (a) shows the (periodic) stencil $[1\ -2\ 1]/h^2$ and part (b) the periodic PS matrix. Parts (c) and (d) show

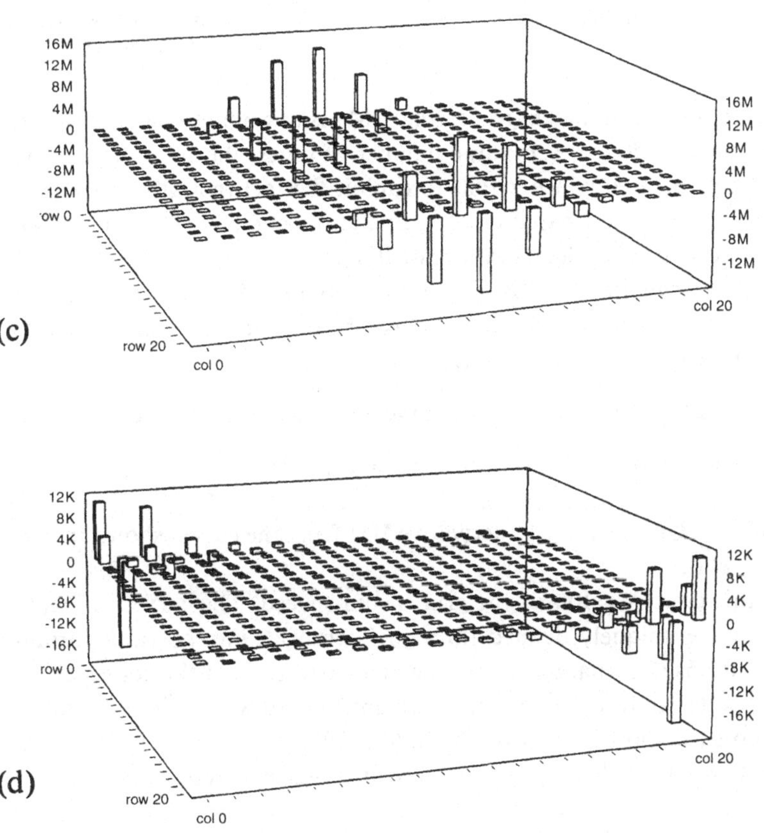

Figure 4.3-1. **(c)** PS, nonperiodic, equi-spaced grid $[13 \times 10^6]$.
(d) PS, nonperiodic, Chebyshev $[17 \times 10^3]$.

the nonperiodic equi-spaced and Chebyshev matrices ($\gamma = 0$ and 0.5, respectively). Large elements are seen in the top and bottom rows (corresponding to approximations near the boundaries).

For nonperiodic problems, the algorithm in Section 3.1 (code in Appendix C) can be used to generate DMs very conveniently. In the PS case:

```
                                               Specify size of grid
      PARAMETER (N=... , M=... )               and highest derivative
      IMPLICIT REAL*8 (A-H,O-Z)
      DIMENSION X(0:N),C(0:N,0:M),DM(0:N,0:N,M)
      DO 10 I=0,N
10       X(I) = ....                           Specify the gridpoints
      DO 20 I=0,N
         CALL WEIGHTS1 (X(I),X(0),N,M,C)
         DO 20 L=1,M
            DO 20 J=0,N
20          DM(I,J,L) = C(J,L)                 DM(*,*,L) contains now the
 :                                             DMs for the Lth derivative,
                                               L = 1, 2, ..., M.
```

The computer time taken generating DMs is seldom critical. However, if this must be done a very large number of times, the code shown here should not be used (since it fails to exploit the fact that all the separate calls to WEIGHTS1 are based on the same grid - some intermediate quantities need not be recalculated repeatedly).

The elements in PS DMs are very closely linked to the terms in Lagrange's interpolation formula. They can be read off from curves such as those shown in Figure 3.3-2 as follows:

$$D^p_{i,j} = p\text{th derivative of curve number } j \text{ at location } x_i. \qquad (4.3\text{-}1)$$

This relation is a consequence of equation (3.1-2).

Example. Generate a first-derivative DM from the curves shown in Figure 3.3-2(a).

In this case, $N = 10$ and the grid is equi-spaced. The curves in Figure 3.3-2(a) are uniquely defined through Lagrange's interpolation formula, equations (3.3-1) and (3.3-2). The first derivatives of the curves at the node points can be calculated in a straightforward way (tediously, unless one spots the shortcut that results in (4.3-2)) and then laid out as shown in Table 4.3-1. In this layout, the derivative values directly form the desired DM.

The top line $(-\frac{7381}{2520}, 10, -\frac{45}{2}, 40, \ldots, -\frac{1}{10})$ would have appeared as line 10 for the first derivative in Table 3.1-2, had that table extended to include order 10. The center line $(-\frac{1}{1260}, \frac{5}{504}, -\frac{5}{84}, \frac{5}{21}, -\frac{5}{6}, 0, \frac{5}{6}, -\frac{5}{21}, \frac{5}{84}, -\frac{5}{504}, \frac{1}{1260})$

Table 4.3-1. *First derivative of the polynomials in Figure 3.3-2(a), evaluated at the node points*

Node location on curve	Curve number 0	1	2	3	4	5	6	7	8	9	10
−1.0	$-\frac{7381}{2520}$	10	$-\frac{45}{2}$	40	$-\frac{105}{2}$	$\frac{252}{5}$	-35	$\frac{120}{7}$	$-\frac{45}{8}$	$\frac{10}{9}$	$-\frac{1}{10}$
−0.8	$-\frac{1}{10}$	$-\frac{4609}{2520}$	$\frac{9}{2}$	-6	7	$-\frac{63}{10}$	$\frac{21}{5}$	-2	$\frac{9}{14}$	$-\frac{1}{8}$	$\frac{1}{90}$
−0.6	$\frac{1}{90}$	$-\frac{2}{9}$	$-\frac{341}{280}$	$\frac{8}{3}$	$-\frac{7}{3}$	$\frac{28}{15}$	$-\frac{7}{6}$	$\frac{8}{15}$	$-\frac{1}{6}$	$\frac{2}{63}$	$-\frac{1}{360}$
−0.4	$-\frac{1}{360}$	$\frac{1}{24}$	$-\frac{3}{8}$	$-\frac{319}{420}$	$\frac{7}{4}$	$-\frac{21}{20}$	$\frac{7}{12}$	$-\frac{1}{4}$	$\frac{3}{40}$	$-\frac{1}{72}$	$\frac{1}{840}$
−0.2	$\frac{1}{840}$	$-\frac{1}{63}$	$\frac{3}{28}$	$-\frac{4}{7}$	$-\frac{11}{30}$	$\frac{6}{5}$	$-\frac{1}{2}$	$\frac{4}{21}$	$-\frac{3}{56}$	$\frac{1}{105}$	$-\frac{1}{1260}$
0.0	$-\frac{1}{1260}$	$\frac{5}{504}$	$-\frac{5}{84}$	$\frac{5}{21}$	$-\frac{5}{6}$	0	$\frac{5}{6}$	$-\frac{5}{21}$	$\frac{5}{84}$	$-\frac{5}{504}$	$\frac{1}{1260}$
0.2	$\frac{1}{1260}$	$-\frac{1}{105}$	$\frac{3}{56}$	$-\frac{4}{21}$	$\frac{1}{2}$	$-\frac{6}{5}$	$\frac{11}{30}$	$\frac{4}{7}$	$-\frac{3}{28}$	$\frac{1}{63}$	$-\frac{1}{840}$
0.4	$-\frac{1}{840}$	$\frac{1}{72}$	$-\frac{3}{40}$	$\frac{1}{4}$	$-\frac{7}{12}$	$\frac{21}{20}$	$-\frac{7}{4}$	$\frac{319}{420}$	$\frac{3}{8}$	$-\frac{1}{24}$	$\frac{1}{360}$
0.6	$\frac{1}{360}$	$-\frac{2}{63}$	$\frac{1}{6}$	$-\frac{8}{15}$	$\frac{7}{6}$	$-\frac{28}{15}$	$\frac{7}{3}$	$-\frac{8}{3}$	$\frac{341}{280}$	$\frac{2}{9}$	$-\frac{1}{90}$
0.8	$-\frac{1}{90}$	$\frac{1}{8}$	$-\frac{9}{14}$	2	$-\frac{21}{5}$	$\frac{63}{10}$	-7	6	$-\frac{9}{2}$	$\frac{4609}{2520}$	$\frac{1}{10}$
1.0	$\frac{1}{10}$	$-\frac{10}{9}$	$\frac{45}{8}$	$-\frac{120}{7}$	35	$-\frac{252}{5}$	$\frac{105}{2}$	-40	$\frac{45}{2}$	-10	$\frac{7381}{2520}$

Notes: Each entry is to be divided by h, where $h = 0.2$. The entries form the nonperiodic $N = 10$ PS DM in the case of the first derivative and an equi-spaced grid. The quantity $1/h = 5$ is factored out in order to simplify a comparison with Tables 3.1-1 and 3.1-2. The top line here, corresponding to the node location −1.0 on the curves in Figure 3.3-2(a), would have appeared as line (= order of accuracy) 10 for the first-derivative case in Table 3.1-2. The central line (node location 0.0 on curves) would similarly have appeared as the first omitted line for the first derivative in Table 3.1-1.

similarly would have appeared as the order-10 approximation for the first derivative in Table 3.1-1.

Nielsen (1956, pp. 150–4) noted how relation (4.3-1) and a clever arrangement of the arithmetic in Lagrange's interpolation formula allowed the elements of D^1 to be computed in only four operations per element (to leading order when N is large):

$$D^1_{jk} = \begin{cases} \dfrac{a_j}{a_k(x_j - x_k)} & \text{if } j \neq k, \\ \displaystyle\sum_{\substack{i=0 \\ i \neq k}}^{N} \frac{1}{x_k - x_i} & \text{if } j = k, \end{cases} \qquad \text{where } a_k = \prod_{\substack{i=0 \\ i \neq k}}^{N} (x_k - x_i). \tag{4.3-2}$$

Table 4.3-2. *Weights for FD approximations of the first derivative as derived from the Haar and Daubechies wavelets*

Wave-let	Order of accuracy	Right half of (antisymmetric) table of weights; $x=$ 0	1	2	3	4	5	6	…	…
H	2	0	$\frac{1}{2}$							
D_4	4	0	$\frac{2}{3}$	$-\frac{1}{12}$						
D_6	6	0	$\frac{272}{365}$	$-\frac{53}{365}$	$\frac{16}{1095}$	$\frac{1}{2920}$				
D_8	8	0	$\frac{39296}{49553}$	$-\frac{76113}{396424}$	$\frac{1664}{49553}$	$-\frac{2645}{1189272}$	$-\frac{128}{743295}$	$\frac{1}{1189272}$		
$\vdots$	$\vdots$	$\vdots$	$\vdots$	$\vdots$	$\vdots$	$\vdots$	$\vdots$	$\vdots$	$\vdots$	$\ddots$

DMs for higher derivatives can in this case be obtained as matrix powers of D^1. A much less costly recursion is also available: D^p can be obtained from D^{p-1} in only five operations per element ($p=2,3,\ldots$; Huang and Sloan 1994, Welfert 1994). Welfert further notes the following.

- The PS literature contains many instances of authors assuming the relation $D^p=(D^1)^p$ when in fact it does not hold (for example, it fails for the periodic PS method if the number of points is even). A sufficient condition for this relation is presented.
- Closed-form expressions for D^1_{jk} and D^2_{jk} become particularly simple for many cases of orthogonal polynomials. A comprehensive list is collected.

Rounding-error propagations within different methods for calculating Chebyshev DMs are discussed by Breuer and Everson (1992).

Most numerical methods that produce approximations to derivatives from discrete data can be represented using DMs and FD weights. (Needless to say, these weights alone convey insufficient information for assessing a method's flexibility or utility.) In the case of approximating first derivatives with use of the Haar wavelet H and the Daubechies wavelets $D_4, D_6, \ldots$, the results become exact for polynomials of increasing order (Beylkin 1992). The weights in Table 4.3-2 can be compared to the ones in Table 3.1-1 (for the first derivative).

4.4. Eigenvalues of differentiation matrices

Example 1. Periodic PS; advection equation $u_t=u_x$.

The DM is derived explicitly both in Sections 2.3 and 3.5 (cf. equations (2.3-4) and (3.5-1)). The DM is antisymmetric. Its eigenvalues (EVs) are equi-spaced on the imaginary axis between $-N(\pi/2)i$ and $+N(\pi/2)i$ (when N is odd; there are very minor differences for N even).

The DMs derived in Section 4.3 for nonperiodic PS approximations are nilpotent; that is, *all* their EVs are equal to zero (a consequence of the fact that a finite number of differentiations of any polynomial leads to an identically zero result). Adding boundary conditions (BCs) normally leads to nonsingular DMs, with *no* eigenvalue being zero. A major difficulty is that some of these EVs can be spurious - and very large. This adversely affects time stepping techniques (to be discussed in Section 4.5). Many of the special techniques in Chapter 5 are designed to (partially) overcome this. In order to provide a background for these upcoming discussions, we next describe some typical EV distributions for nonperiodic problems.

Example 2. Chebyshev PS; advection equation $u_t = u_x$, $u(1) = 0$.

Figure 4.4-1 shows the EVs of the DM incorporating $u(1) = 0$ for $N = 8, 16, 32$, and 64. Although most of the EVs converge to a curve in the left half-plane between $-iN$ and $+iN$, a few spurious outliers diverge at rates proportional to N^2.

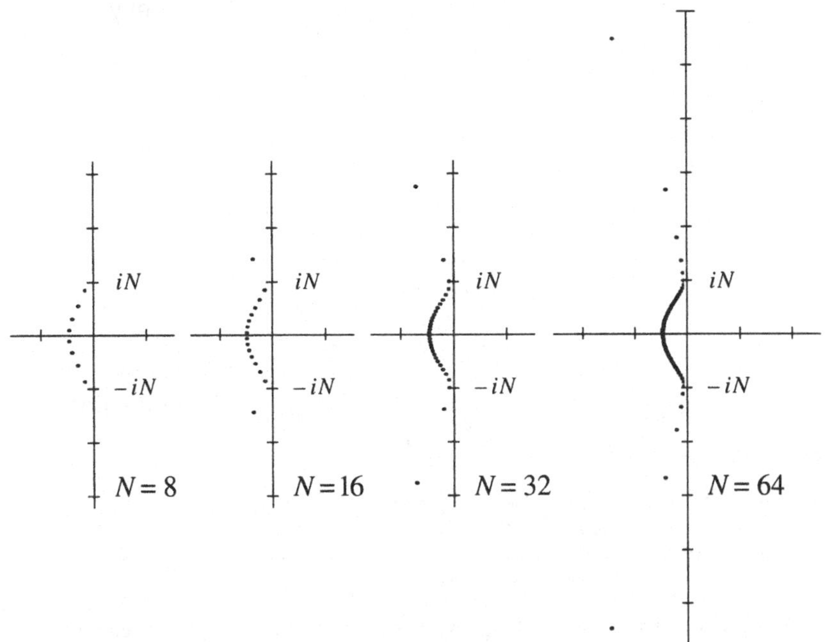

Figure 4.4-1. Eigenvalues of Chebyshev DMs for an advection equation.

Trefethen and Trummer (1987) note that the small (physical) eigenvalues for large values of N are very sensitive to rounding errors (curiously, such errors leave them distributed along different distinct paths).

When the gridpoints are instead distributed as the zeros of Legendre polynomials $P_N(x)$, Dubiner (1987) noted that the spurious outliers were absent in the particular model problem. However, the EV sensitivity remains large (and EVs alone fail to fully describe stability issues in cases such as this, where the DMs are highly nonnormal matrices). It is questionable whether the use of Legendre polynomials offers any practical advantage (Trefethen 1988).

Example 3. Chebyshev and equi-spaced nonperiodic PS approximation of the heat equation $u_t = u_{xx}$, $u(\pm 1) = 0$.

The continuous problem $u_{xx} = \lambda u$, $u(\pm 1) = 0$, has the eigenvalues $\lambda_k = -(k\pi/2)^2$, $k = 1, 2, \ldots$. The EVs of the Chebyshev DM for u_{xx} are all real and negative (Gottlieb and Lustman 1983; their key observation in proving this was that the coefficients in the characteristic polynomial can be found analytically). In the case of $N = 20$, this DM is the matrix in Figure 4.3-1(d) with the first and last rows and columns removed. Figure

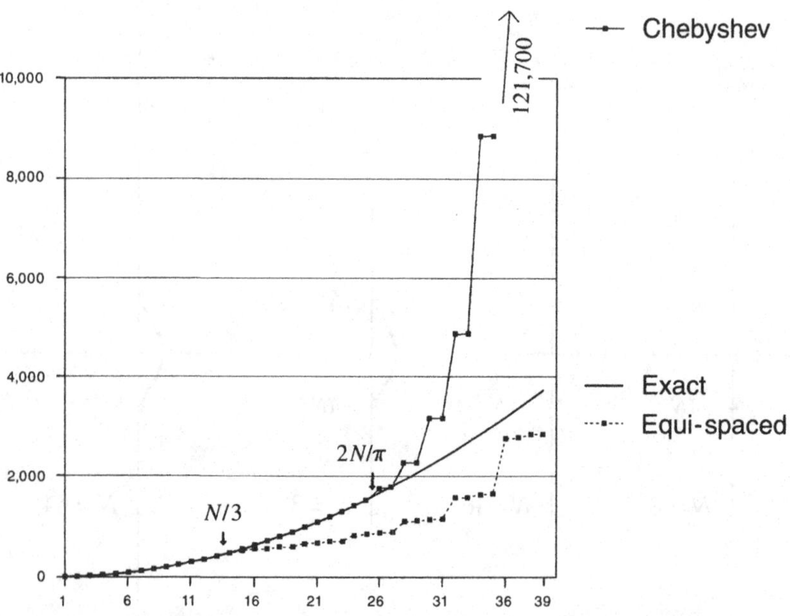

Figure 4.4-2. Magnitudes of eigenvalues of PS DMs (for equi-spaced and Chebyshev grids) compared to analytic eigenvalues; $N = 40$.

4.4-2 compares the magnitude of its EVs and those for the equi-spaced nonperiodic PS method (cf. Figure 4.2-1(c); in this case, many higher EVs are complex) with the EV magnitudes for the continuous problem. The portions $2/\pi$ and 1/3 of the EVs are spectrally accurate in the two cases (cf. the numbers 2, 6, and π of points per wavelength mentioned in Section 4.1).

Pseudospectral methods can also be devised for infinite domains (this is discussed further at the end of Section 5.5). For eigenvalues of Hermite and "rational spectral" PS DMs, see Weideman (1992). Other infinite domain PS DM studies concern Laguerre eigenvalues (Funaro 1992) and sinc eigenvalues (Stenger 1993).

4.5. Time stepping methods and stability conditions

The examples in this section will incorporate one space direction and time. The discrete step sizes are denoted h and k, respectively. We call a scheme *stable* if its numerical solution does not diverge at a fixed time T when h and $k \to 0$ (often, this will require certain relations to hold between h and k).

Time stepping methods

Several different approaches are available for time stepping FD and PS methods.

Method of lines (MOL). We discretize in space only and apply a "packaged" ODE solver to advance the discrete system in time.

Many numerical ODE solution techniques have been thoroughly analyzed, and convenient program packages have been developed. The user need not be concerned with many tedious issues such as starting techniques for multi-step methods, time step and order adjustments, and so forth.

In its very simplest form, the MOL time stepping could consist of a fixed time step and fixed-order implementation of, say, leapfrog or a Runge–Kutta method that the user incorporates "in-line" in the code (instead of invoking a library routine).

The ODE methods that are used are either *explicit* or *implicit* – emphasizing speed per time step or large stability domains, respectively. These methods are discussed briefly in this chapter and with more detail in Appendix G.

Fractional step. We separate the terms in the RHS (containing space derivatives, forcing functions, etc.) into m groups, and then advance $m\times$ {each group} in turn a time distance of k/m using (different) one-step methods. This approach has significant strengths.

+ It allows separation of multi-dimensional space operators into 1-D operators – the ADI (alternative direction implicit) concept. Linear and nonlinear terms can be separated to allow the use of highly effective specialized time stepping methods for each fractional step.
+ Stability analysis can be very simple: It suffices to find some norm in which the solution does not grow too fast during any of the fractional steps.

The fractional step approach also involves some complications.

- Implementation of boundary conditions – it is often unclear how to decompose these to suit the different characters of the separate equations.
- The time accuracy is limited to second order (however, improvements are possible through Richardson-type extrapolation).

  To achieve even second order, the first and last split steps may need to differ somewhat from the intermediate ones.

An example of fractional time stepping is given in Section 8.2. More extensive discussions can be found for example in Boyd (1989, Chap. 10).

Combined time-space stencils. In these, a relatively large time discretization error of a simple stencil has been reduced by also invoking a differentiated form of the governing equations.

Example 1. Devise a forward Euler–FD2-type stable, second-order scheme for the equation $u_t+u_x=0$.

The immediate forward Euler–FD2 approximation

$$\frac{u(x,t+k)-u(x,t)}{k}=-\frac{u(x+h,t)-u(x-h,t)}{2h}$$

is only first-order accurate in time (k), and is also unconditionally unstable. Its leading error term $(k/2)u_{tt}$ is equal to $(k/2)u_{xx}$. Explicitly subtracting this out leads to the second-order accurate and conditionally stable Lax–Wendroff scheme

$$\frac{u(x,t+k)-u(x,t)}{k}=-\frac{u(x+h,t)-u(x-h,t)}{2h}$$
$$+\frac{k(u(x+h,t)-2u(x,t)+u(x-h,t))}{2h^2}.$$

A convenient two-step variation of this for systems of conservation laws is described in Richtmyer and Morton (1967, pp. 302–3).

This approach of exchanging time errors for additional space terms appears to be used only rarely in connection with PS methods, possibly because PS methods are not always viewed as special cases of FD methods.

Specialized techniques can exploit particular features of different classes of equations. For instance, three of the methods for the Korteweg–de Vries equation that are given in Section 8.2 exploit the fact that the highest-order derivative is linear and without variable coefficient. Another example is offered by a spectrally accurate time stepping scheme of Tal-Ezer, Carcione, and Kosloff (1990) for linear problems with time-independent variable coefficients (this is applicable e.g. to the seismic modeling problem described in Section 8.4).

Time stepping methods are usually selected to achieve some prescribed accuracy at the lowest cost – that is, to minimize the product of

- the cost per time step and
- the number of time steps needed.

The former is usually easy to estimate. Useful quantities for assessing the latter include:

Convergence rate (for the time stepping part) – typically $O(k^p)$, where p is some integer.

Stability condition – if present, often in the form $k/h^q <$ const., where q is a (low) integer.

Consistency – simply a requirement that *locally,* the FD stencil approximates the governing equation (with an error that decreases as h and k tend to zero).

The following key result connects these three quantities.

Lax equivalence theorem. *For a well-posed linear problem, a consistent approximation converges if and only if it is stable.*

In other words, as the mesh is refined, either the numerical solution will diverge (which would be very apparent in a calculation) or it will converge to the proper solution; it can neither simply drift nor settle toward anything else. By analyzing stability, the convergence/divergence status for a scheme can be settled before it is implemented numerically.

The convergence rates p for time stepping schemes are well known (and independent of spatial approximation methods). This leaves stability as the key issue to check when assessing the utility of different schemes, that is, of different combinations of time and space discretizations.

Stability conditions

Stability analysis for FD approximations to PDEs almost got off to an early start in the pioneering paper by Richardson (1910). At that time, FD methods were extremely novel; he notes:

Step-by-step arithmetical methods of solving ordinary difference equations have long been employed for the calculation of interest and annuities. Recently their application to differential equations has been very greatly improved by the introduction of rules allied to those for approximate quadrature.

The main topic of the paper is a beautifully executed FD calculation of the stresses in the 1902 dam across the Nile at Asswan (a 2-D calculation – in an irregular geometry and requiring local mesh refinement – that was prompted by the failure of a similar dam at Bouzey, France). Before addressing that topic, Richardson introduced the concept of FD schemes through some examples. One of these was the following.

Example 2. Use the leapfrog approximation

$$\frac{u(x,t+k)-u(x,t-k)}{2k}=\frac{u(x-h,t)-2u(x,t)+u(x+h,t)}{h^2}$$

to time step the heat equation $\partial u/\partial t=\partial^2 u/\partial x^2$, with initial condition $u(x,0)=1,\ -\frac{1}{2}<x<\frac{1}{2}$, and boundary conditions $u(\pm\frac{1}{2},t)=0,\ t\geq 0$.

Richardson exploited the symmetry at $x=0$ and used a special (rather inaccurate) starting procedure to obtain values at $t=0.001$. He calculated five time steps ($h=0.1$, $k=0.001$) and obtained the values shown in Table 4.5-1. As Figure 4.5-1 confirms, the errors do initially decrease.

Table 4.5-1. *Results of five time steps in Richardson's example (copied from his paper)*

	t						$t=$	
x	0	0.001	0.002	0.003	0.004	0.005	0.005[a]	Errors
0.5	0.0000	0.0000	0.0000	0.0000	0.0000	0.0000	0.0000	
0.4	1.0000	0.9090	0.8356	0.7714	0.7209	0.6729	0.6828	−0.0099
0.3	1.0000	0.9959	0.9834	0.9695	0.9492	0.9329	0.9545	−0.0216
0.2	1.0000	0.9998	0.9993	0.9968	0.9945	0.9887	0.9980	−0.0093
0.1	1.0000	1.0000	1.0000	0.9999	0.9994	0.9990	0.9996	−0.0006
0	1.0000	1.0000	1.0000	1.0000	1.0000	0.9998	1.0000	−0.0002

[a] Numbers in this column are taken from the analytic solution.

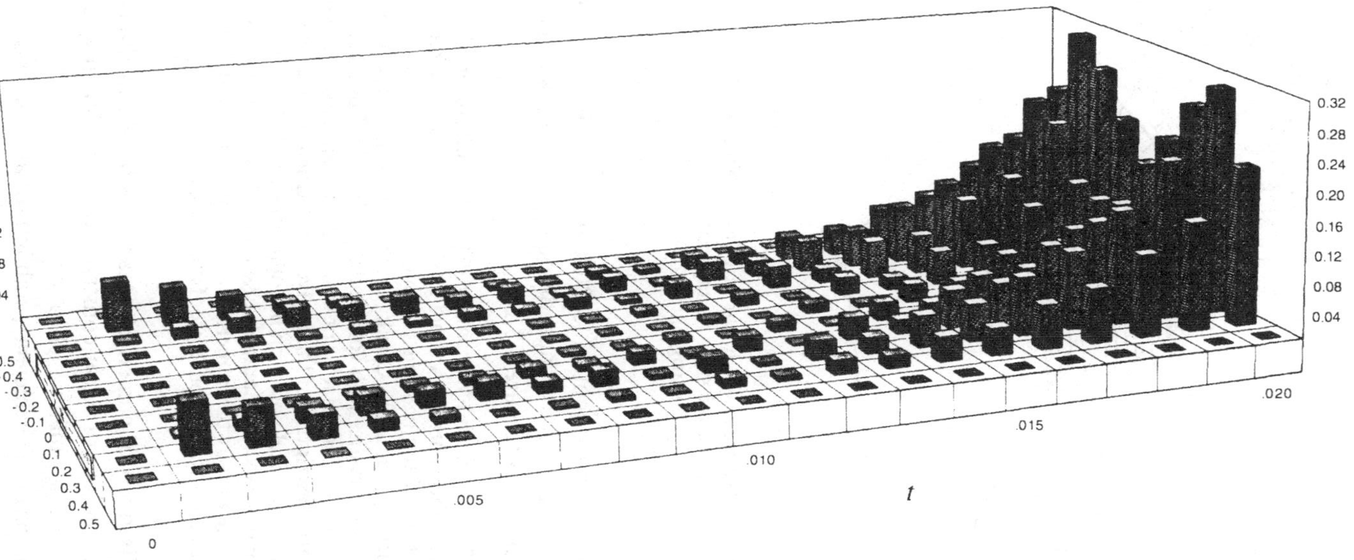

Figure 4.5-1. Magnitude of the error at the gridpoints during the first 20 time steps in the example by Richardson (leapfrog time stepping of the heat equation).

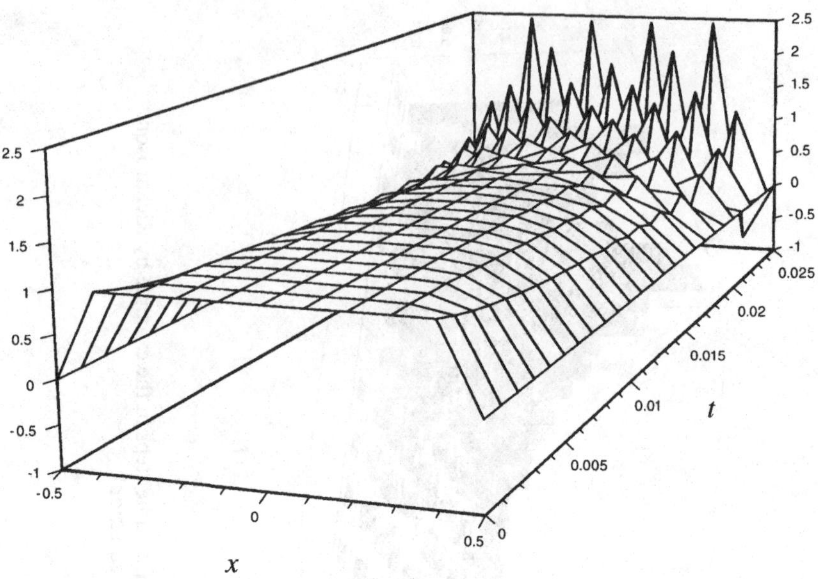

Figure 4.5-2. Typical structure of a linear instability. Numerical solution of the Richardson example, run 25 time steps (compared to 20 in Figure 4.5-1 and 5 in his original study).

However, had he continued another 15–20 steps, he would have seen a very different pattern. The scheme is unconditionally unstable, and its solution breaks up in oscillations (Figure 4.5-2).

As late as 1937 (Hartree and Womersley 1937), no fundamental flaw had been found with this scheme. Around 1945, Hartree's co-worker Phyllis Nicolson switched (after much frustration on a problem related to burning wood) to a scheme proposed by Crank (Crank and Nicolson 1947). Hartree corresponded on the matter with von Neumann, who (at Los Alamos during the war) had developed what we now describe as von Neumann stability analysis. An early description of this is given in O'Brien, Hyman, and Kaplan (1951). Some historical notes can be found in Bell (1986).

The now-famous paper by Courant, Friedrichs, and Lewy (1928) was not much noted at the time by the numerical community, since it discussed FD schemes only as tools for existence proofs (cf. Lax 1967).

Following the discovery of numerical instabilities for FD methods, it was soon found that analysis of linear constant-coefficient initial value problems was insufficient for many applications.

1. In cases of variable coefficients, stability for all problems with "frozen" coefficients is neither necessary nor sufficient for stability (Kreiss 1962, Richtmyer and Morton 1967).

However, such examples are not typical. For large classes of problems, testing with frozen coefficients is appropriate to decide both well-posedness of the governing equations and stability of FD schemes.

2. With boundaries present, local mode analysis for the interior may not be sufficient even for constant-coefficient problems. For FD schemes of fixed stencil width, GKS (Gustafsson, Kreiss, and Sundström 1972) analysis addresses wave reflection at boundaries (see also Kreiss 1968, 1970). This analysis provides necessary and sufficient conditions for the stability of initial boundary value problems. Simplified (analytic) versions are described by Trefethen (1983) and Goldberg and Tadmor (1985). Thuné (1990) describes a numerical procedure for stability verification. The boundary instability shown in part (b) of Figure 4.5-3 is seen to develop through several stages in which both boundaries play a role.

Slight amounts of numerical dissipation (the damping of high modes that occurs naturally in most ODE solvers, as seen in the stability diagrams in Appendix G) greatly reduce the threat of instabilities caused by spurious waves bouncing between boundaries.

3. Nonlinear instabilities can arise independently of the time stepping scheme that is used. Such instabilities are discussed in Section 4.6.

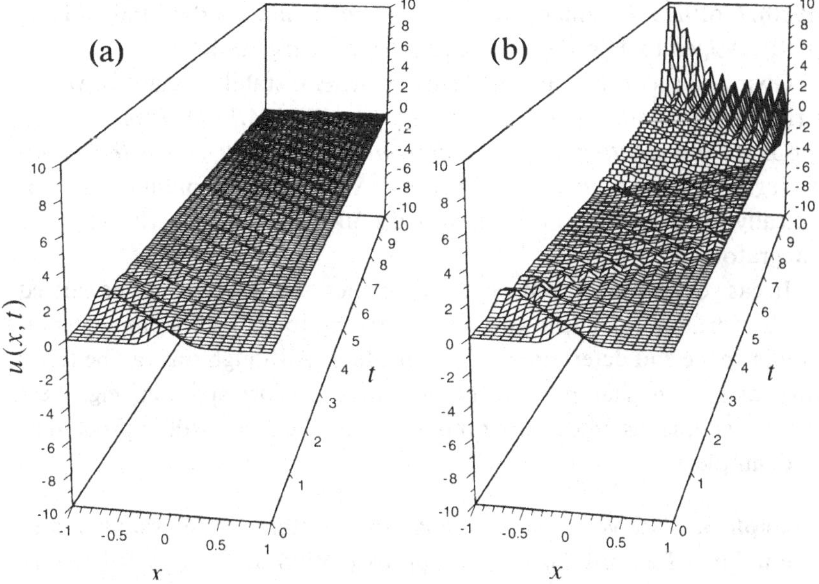

Figure 4.5-3. Leapfrog–FD2 solutions of $u_t = u_x$ displayed at all gridpoints ($h = 0.1$, $k = 0.09$). The left boundary condition was **(a)** $u(-1,t) = u(-1+h, t-k)$ and **(b)** $u(-1,t) = u(-1+h,t)$; the right boundary condition was in both cases $u(1,t) = 0$.

Even numerical schemes that are constrained by conservation laws can break down into seemingly random noise. This problem arises quite frequently for nonlinear wave equations, and is mentioned briefly in Section 8.2.

Gustafsson, Kreiss, and Oliger (1995) survey issues such as well-posedness of PDEs and FD stability tests.

Additional difficulties concerning stability that are more specific to PS methods include the following.

4. One needs to distinguish between *Lax* stability (fixed T with h and $k \to 0$) and *eigenvalue* stability (fixed h and k as $T \to \infty$). For the highly nonnormal DMs that arise from nonperiodic PS methods, very large growths can arise initially even in cases when eigenvalue analysis suggests stability (cf. the Kreiss matrix theorem in Richtmyer and Morton 1967). Reddy and Trefethen (1990) use "pseudospectra" to provide further insight into this and similar phenomena.

5. PS methods can be unstable even when the corresponding FD methods of increasing orders *all* are stable. Tadmor (1987) addresses this phenomenon in connection with a linear model equation $u_t = c(x)u_x$.

Energy methods provide a particularly powerful general tool for PS stability analysis. Gustafsson et al. (1995, Chap. 11) discuss this technique for FD methods; for nonperiodic PS methods, see Gottlieb and Turkel (1980), Gottlieb, Lustman, and Tadmor (1987), and Gottlieb and Tadmor (1991). Appendix H gives a few examples of energy estimates.

Owing to the technical complexity of general stability analysis, we restrict the discussion here to *eigenvalue stability*: *All the eigenvalues of the DMs representing the space derivatives must fall within the stability region for the time stepping method.* Although this condition is theoretically insufficient, it provides very useful guidelines for selecting time integrators.

In case of constant-coefficient FD schemes, this analysis can be carried out by forming the cyclic DM corresponding to a periodic implementation in space and determining its eigenvalues. Although this can be fairly easy (as seen in Example 5 to follow, where the corresponding eigenvectors are simple discrete Fourier modes), it can be easier still to proceed as in Example 3.

Example 3. Use von Neumann analysis to determine the stability condition for a forward Euler–FD2 approximation and a forward Euler–periodic PS approximation to the heat equation $\partial u/\partial t = \partial^2 u/\partial x^2$.

FD2: Substituting $u(x,t) = \xi^{t/k} e^{i\omega x}$ into the FD stencil

$$\frac{u(x,t+k)-u(x,t)}{k}=\frac{u(x-h,t)-2u(x,t)+u(x+h,t)}{h^2}$$

gives (after minor simplification)

$$\frac{\xi-1}{k}=\frac{e^{-i\omega h}-2+e^{i\omega h}}{h^2}; \tag{4.5-1}$$

that is, $\xi=1-4k/h^2\sin^2(\omega h/2)$. Stability requires $|\xi|\le 1+O(k)$ for all ω that can be represented on the grid; $|\omega|\le \pi/h$ (with the extreme modes $\omega=\pm\pi/h$ corresponding to the sawtooth pattern ..., +1, −1, +1, −1, +1, −1, ...). The requirement on $|\xi|$ ensures that $|\xi|^{T/k}$, the growth factor when advancing a time T in steps of size k, remains bounded as $k\to 0$. Here, this yields $k/h^2\le\frac{1}{2}+O(k)$. Omitting the $O(k)$ term (as we will in the following discussions), we obtain the requirement for no spurious growth of any Fourier mode: $k/h^2\le\frac{1}{2}$.

Fourier–PS: The RHS of (4.5-1) is the factor by which $e^{i\omega x}$ is multiplied when the second-derivative approximation FD2 is applied. For the Fourier–PS method, this factor is instead $-\omega^2$; that is, $\xi=1-k\omega^2$. Recalling that $|\omega|\le\pi/h$, the stability requirement $|\xi|\le 1$ becomes $k/h^2\le 2/\pi^2$ (≈ 0.2026).

Stability regions

An ODE solver is *stable* for step size k and (complex) eigenvalue λ if the numerical solution to $u_t=\lambda u$ does not grow with t. It is called *A-stable* if it is stable for all λ in the negative half-plane (this is the ideal situation, matching this same property of the analytic solution $u(t)=e^{\lambda t}$). For a major class of ODE solvers, A-stability is not possible.

Second Dahlquist stability barrier (Dahlquist 1956, 1985). For linear multi-step methods (see Section G.2):

- an explicit method cannot be A-stable; and
- an implicit A-stable method can be second-order accurate ($p=2$) at best.

Weaker requirements than A-stability are often sufficient in applications. Most methods (e.g. all explicit linear multi-step or Runge–Kutta methods) have stability domains that are bounded in all directions away from the origin.

Example 4. Determine the stability domain for forward Euler time stepping.

The forward Euler approximation to $u'=\lambda u$ becomes $u(t+k)=u(t)+k\lambda u(t)$, that is, $u(t+nk)=(1+\lambda k)^n u(t)$. The condition for no growth, $|1+\lambda k|\le 1$, implies that the stability region (values of λk satisfying this inequality) is a circle with radius 1 centered at -1.

> Very low accuracy and no stability coverage along the imaginary axis render the forward Euler approach very unattractive in general, and unconditionally unstable for wave-type problems.

Commonly used ODE solvers represent compromises between low operation counts, high accuracies, and large stability domains. They include many Runge–Kutta (RK) schemes, Adams-type methods, and – for "stiff" problems, with some eigenvalues close and others far away in the left half-plane – backward differentiation (BDF) methods. For discussions on ODE solvers, see for example Gear (1971), Shampine and Gordon (1975), Hairer, Nørsett, and Wanner (1987), Hairer and Wanner (1991), and Lambert (1991). Many stability domains are illustrated in Sand and Østerby (1979). Illustrations for a few of the most commonly used methods are given in Appendix G.

Example 5. Determine the stability conditions for solving $\partial u/\partial t+\partial u/\partial x=0$ with the fourth-order Runge–Kutta procedure in time, and with different centered FD methods (including Fourier–PS) in space.

The EVs λ of the PS DM lie on the imaginary axis:

$$\lambda\in i[-N(\pi/2),N(\pi/2)]=i[-\pi/h,\pi/h]$$

(cf. Example 1 in Section 4.4). The same fact is displayed as the range of the PS curve in Figure 4.1-3; this figure shows also that the corresponding λ ranges (omitting $\pm i$) for FD2 and FD4 extend to $1/h$ and approximately $1.3722/h$, respectively.

The stability domain for RK4 extends from the origin a distance of $2\sqrt{2}$ up and down the imaginary axis (see Appendix G and Figure G.1-1). The stability conditions therefore become $|\lambda k|\le 2\sqrt{2}$. Combining this with the λ ranges for the different spatial approximations gives the results shown in Table 4.5-2. The process of increasing the formal order of accuracy for the first derivative from two to infinity (FD2 to Fourier–PS) reduces the stability condition only by a factor of π. For the same spatial accuracy, h can normally be made much more than π times larger. The stability condition therefore becomes increasingly less restrictive when higher-order methods are used in space.

Table 4.5-2. *Stability conditions for RK4 time stepping of $u_t + u_x = 0$*

Spatial method	Stability condition; upper limit on k/h	Approximate value
FD2	$2\sqrt{2}$	2.828
FD4	$\dfrac{4\sqrt[4]{3}}{(1+\sqrt{6})\sqrt{\sqrt{2}-\frac{1}{2}\sqrt{3}}}$	2.061
FD6	—	1.783
FD8	—	1.634
⋮	⋮	⋮
Fourier-PS	$\dfrac{2\sqrt{2}}{\pi}$	0.900

Dissipation

The stability of an FD scheme can often be improved by adding some high-order dissipation, a damping of high frequency modes that does not affect the order of accuracy. As an example, to the equi-spaced FD4 approximation

$$[\ \tfrac{1}{12}\ \ -\tfrac{2}{3}\ \ 0\ \ \tfrac{2}{3}\ \ -\tfrac{1}{12}\ ]\ \ u = hu' + O(h^5)$$

we can add

$$\beta\cdot[\,1\ \ -6\ \ 15\ \ -20\ \ 15\ \ -6\ \ 1\,]u = \qquad O(h^6),$$

with weights taken from the equi-spaced FD2 approximation to d^6u/dx^6. For a periodic problem, the DM eigenvectors are unchanged (pure Fourier modes). However, with $\beta > 0$, all the eigenvalues (except for the zero one) have been shifted from the imaginary axis into the left half-plane.

In terms of the operators D_+ and D_- (introduced in the derivation of equation (4.1-4)), the most commonly used dissipative terms on equi-spaced grids can be expressed as $\beta\cdot(h^2D_+D_-)^m$, where m is some small integer ($m = 3$ in our example). In the context of advancing a PDE in time, a dissipative term:

- Leaves the order of the truncation error unaffected.

  The error that is introduced is $O(h^{2m})$, and m can be selected high enough so as not to influence the approximation order of the governing equation.

- Damps high frequencies.

For the equation $\partial u/\partial t = \beta(h^2 D_+ D_-)^m$ (with RHS $= O(h^{2m})$), the amplitude $\alpha(t)$ of a Fourier mode $e^{i\omega x}$ satisfies

$$a(t) = a(0)\exp\left[\beta(-1)^m 2^{m+1}\left(\sin\frac{\omega h}{2}\right)^{2m}\right]t.$$

Selecting β such that $\beta(-1)^m < 0$, $a(t)$ will decay for all $\omega \neq 0$, and especially rapidly for the highest frequencies on the grid (recall from Section 4.1 that with grid spacing h, $|\omega| \leq \pi/h$).

Especially for variable-coefficient initial boundary value problems in more than one dimension, formal stability analysis can become exceedingly difficult. Fortunately, stability failures tend to show up clearly and then display some distinctive structure. This makes it often easy to apply an appropriate amount of dissipation at the critical locations.

The technique of introducing a free parameter (like β previously) and then assigning it a specific value can be used also for purposes other than improving stability at domain interiors or boundaries. Examples include:

- creating FD schemes that are not only accurate in general but are also exact in certain situations of particular significance (e.g., satisfying some asymptotic solution at a singular point);
- generating FD stencils with particular entries at certain locations – the example in Table 4.5-3 is illustrative (but not of critical importance, since the algorithm in Section 3.1 generates such stencils directly – and without requiring the grid to be equi-spaced); and

Table 4.5-3. *Illustration of how equi-spaced partly and fully one-sided FD4 approximations can be obtained from the centered one*

	[$\frac{1}{12}$	$-\frac{2}{3}$	0	$\frac{2}{3}$	$-\frac{1}{12}$]			*$hu'+O(h^5)$ at center location* Same centered stencil as shown in Table 3.1-1.
$-\frac{1}{12}\cdot$	[1	-5	10	-10	5	-1]		*$O(h^5)$ at all locations* Choose the multiple so that the leftmost entry is eliminated when adding.
$=$	0	[$-\frac{1}{4}$	$-\frac{5}{6}$	$\frac{3}{2}$	$-\frac{1}{2}$	$\frac{1}{12}$]		*$hu'+O(h^5)$ at next-to-leftmost grid position*
$+\frac{1}{4}\cdot$		[1	-5	10	-10	5	-1]	Add another multiple to eliminate the leftmost entry.
$=$	0	0	[$-\frac{25}{12}$	4	-3	$\frac{4}{3}$	$-\frac{1}{4}$]	*$hu'+O(h^5)$ at leftmost grid position* Same one-sided stencil as shown in Table 3.1-2.

Notes: The weights for the stencil that is added at each stage are taken from the most compact (lowest-order) approximation possible for d^5u/dx^5. All approximations are $O(h^5)$ at the position that is located inside the ruled rectangle.

- generating high-order compact approximations to differential equations. As an example, the well-known *Mehrstellenverfahren* by Collatz (1960) for the equation $u'' = f$ can be derived as follows:

$$[-\tfrac{1}{12}\ \tfrac{4}{3}\ -\tfrac{5}{2}\ \tfrac{4}{3}\ -\tfrac{1}{12}]/h^2 u = f + O(h^4).$$

By differentiation, $u^{(4)} = f''$. Therefore, to second order:

$$[1\ -4\ 6\ -4\ 1]/h^4 u = [1\ -2\ 1]/h^2 f + O(h^2).$$

Adding $\beta = h^2/12$ of the latter stencil to the former removes the "outliers", and results in the compact fourth-order approximation

$$[1\ -2\ 1]/h^2 u = \tfrac{1}{12}[1\ 10\ 1] f + O(h^4).$$

This approach is easily generalized to many equations in higher dimensions, with variable coefficients, etc.

4.6. Aliasing and nonlinear instabilities

A Fourier expansion of a continuous function typically contains an infinite number of frequencies:

$$u(x) = \sum_{\omega=-\infty}^{\infty} \hat{u}(\omega) e^{i\pi\omega x}.$$

On a discrete set of N equi-distant points in the interval $[-1,1]$, all Fourier modes $e^{i\pi(\omega+kN)x}$, $k = \ldots, -2, -1, 0, 1, 2, \ldots$, look identical. Figure 4.6-1 illustrates this (for the imaginary part) in the case of $N = 10$, $\omega = 1$, and $k = -1, 0$. A discrete Fourier transform (DFT), sampling the function $u(x)$ at such equi-spaced points, cannot distinguish between these modes. To keep the notation simple, we assume N to be odd: $N = 2m+1$. Each discrete Fourier coefficient $\tilde{u}(\omega)$, $\omega \in [-m, m]$, is the sum of all coefficients with modes that look equivalent on the gridpoints:

$$\tilde{u}(\omega) = \sum_{k=-\infty}^{\infty} \hat{u}(\omega + kN).$$

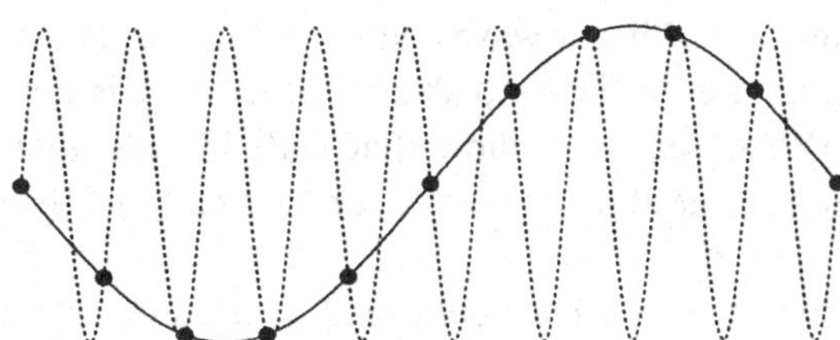

Figure 4.6-1. "Aliasing" - the functions $\sin \pi x$ and $-\sin 9\pi x$ cannot be distinguished when sampled on the equi-spaced grid $x_i = -1 + 0.2i$, $i = 0, 1, \ldots, 10$.

The interpolating trigonometric polynomial (of lowest order, as used in PS approximations)

$$I_N u(x) = \sum_{\omega=-m}^{m} \tilde{u}(\omega) e^{i\pi\omega x}$$

takes the exact values at the gridpoints, but not in between these points. It differs from the truncated Fourier expansion

$$T_N u(x) = \sum_{\omega=-m}^{m} \hat{u}(\omega) e^{i\pi\omega x}$$

by a quantity that has become known as the *aliasing error*:

$$R_N u(x) = I_N u(x) - T_N u(x) = \sum_{\omega=-m}^{m} \sum_{\substack{k=-\infty \\ k\neq 0}}^{\infty} \hat{u}(\omega+kN) e^{i\pi\omega x}.$$

The aliasing error $R_N u(x)$ contains only Fourier modes $|\omega| \leq m$ whereas $u(x) - T_N u(x)$ contains only modes $|\omega| > m$; these quantities are *orthogonal*: $\int_{-1}^{1} R_N u(x) \cdot (u(x) - T_N u(x))\, dx = 0$. Squaring and integrating the equation $u(x) - I_N u(x) = (u(x) - T_N u(x)) - (R_N u(x))$ leads to

$$\|u(x) - I_N u(x)\|^2 = \|u(x) - T_N u(x)\|^2 + \|R_N u(x)\|^2.$$

In this standard L^2 norm, the interpolation error is therefore always larger than the truncation error (unless both are zero).

In the PS method we use the derivatives of $I_N(x)$, rather than of $T_N(x)$, at the gridpoints. Although the size of the difference $R_N u(x)$ can be estimated easily enough (when $u(x)$ is analytic, it decays exponentially fast with increasing N), there is some controversy about how these errors accumulate during a time integration. This is especially the case when complications are present such as irregularities in $u(x)$, low values of N, variable coefficients, nonlinearities, and so on.

Practical experience with PS methods seems very favorable. Aliasing errors are most often oscillatory and therefore tend to cancel out in time-dependent problems. Kreiss and Oliger (1979) prove a result to this effect for the one-dimensional linear wave equation. For steady-state problems also, some proofs are available to show that aliasing is not a problem. Canuto et al. (1988, Sec. 11.3) show that Galerkin (de-aliased) and PS (with aliasing) converge at the same rate for the steady 3-D Navier–Stokes equations.

High-frequency modes (that could lead to aliasing errors) can arise numerically through variable-coefficient terms (e.g. $a(x)(\partial u/\partial x)$) or nonlinear terms (e.g. $u(\partial u/\partial x)$). If $a(x)$ and $u(x)$ contain high Fourier modes then the product will contain much higher modes still, which might then

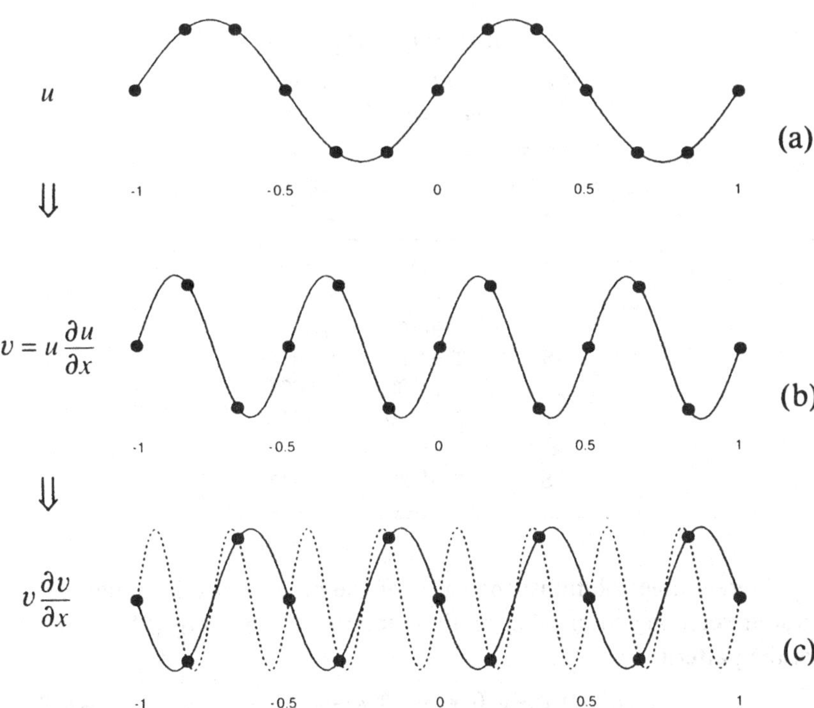

Figure 4.6-2. Frequency doubling (and aliasing) when $u(\partial u/\partial x)$ is formed from a Fourier mode u.

be misinterpreted as lower modes. For example, if $u(x) = \sin \alpha x$ then $u(\partial u/\partial x) = (\alpha/2)\sin 2\alpha x$, a frequency doubling. Figure 4.6-2 illustrates how a moderately high Fourier mode in two steps can generate a mode that will be misinterpreted by grid-based methods.

Numerical nonlinear instabilities

Even when linearized stability conditions are satisfied, numerical schemes can sometimes generate explosively growing (unphysical) spikes. In the pioneering paper in this field, Phillips (1959) finds a 2-D periodic pattern that diverges exponentially fast when a leapfrog time step tends to zero for an FD2 approximation to the incompressible Navier–Stokes equations.

Many later studies on nonlinear instabilities for approximations to convection-dominated flows have focused on the simpler 1-D model equation

$$\frac{\partial u}{\partial t} + \frac{\theta}{2}\frac{\partial u^2}{\partial x} + (1-\theta)u\frac{\partial u}{\partial x} = 0. \qquad (4.6\text{-}1)$$

Table 4.6-1. *Effect of spatial FD order on the rate of a nonlinear instability*

FD order p	c_p	Approximation
2	1/2	0.5
4	3/4	0.75
6	9/10	0.9
8	279/280	0.9964
10	297/280	1.0607
12	243/220	1.1045
⋮	⋮	⋮
PS	$2\pi\sqrt{3}/9$	1.2092

For differentiable solutions, the value of the parameter θ is immaterial. Without referring to aliasing or Fourier modes, Fornberg (1973) notes that the pattern

$$\ldots\ 0\ \epsilon\ -\epsilon\ 0\ \epsilon\ -\epsilon\ 0\ \epsilon\ -\epsilon\ 0\ \ldots \tag{4.6-2}$$

(spatial period $3h$, $\epsilon = \epsilon(t)$) satisfies (4.6-1) for all centered FD approximations to $\partial/\partial x$ (this includes the Fourier–PS method). This pattern is the same one that in Figure 4.6-2 was seen to be capable of generating a multiple of itself – a dangerous feedback possibility. Discretizing in space only, the amplitude $\epsilon(t)$ will satisfy

$$\frac{d\epsilon(t)}{dt} = \frac{c_p(1-3\theta/2)}{h}\epsilon^2(t), \tag{4.6-3}$$

where c_p depends on the order of the FD scheme as shown in Table 4.6-1. Depending on the value of the parameter θ, two different cases arise.

Case 1: $\theta \neq \frac{2}{3}$. Equation (4.6-3) possesses solutions that become infinite after a short time. Irrespective of what scheme is used to discretize in time, letting its time step decrease to zero clearly offers no help against instabilities of this kind.

Case 2: $\theta = \frac{2}{3}$. For all centered FD schemes in space and any spatial period, $(d/dt)\sum_i u_i^2 = 0$; that is, spatially discrete solutions cannot grow in the L^2 norm. However, the computational cost is higher than for $\theta = 0$ or $\theta = 1$.

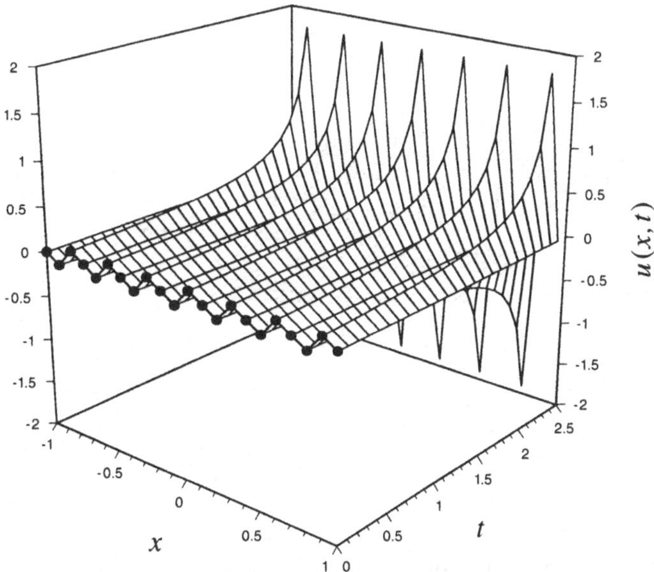

Figure 4.6-3. Nonlinear growth of the pattern (4.6-2) with $\epsilon = -0.1$ in the FD4 solution of $\partial u/\partial t + \frac{1}{2}(\partial u^2/\partial x) = 0$.

Certain time stepping schemes can cause the pattern (4.6-2) to diverge even in the case of $\theta = \frac{2}{3}$. In the case of leapfrog, this was observed in Kreiss and Oliger (1972) and analyzed in Aoyagi (1995).

Period-$2h$ patterns must take the form $\dots\epsilon\,\eta\,\epsilon\,\eta\,\epsilon\,\eta\dots$; these remain constant in time. Richtmyer and Morton (1967) consider a period-$4h$ pattern for $\theta = 1$ whose FD2 approximation diverges when time is stepped by leapfrog. However, experiments show that numerical instabilities for (4.6-1) almost invariably amount to local variations of the period-$3h$ patterm (4.6-2).

Figure 4.6-3 shows the growth of the pattern (4.6-2) in the case of $\epsilon = -0.1$, $h = 2/21$, and $\theta = 1$ when using FD4 in the (periodic) x direction. The analytic solution $\epsilon(t) = -16/(160-63t)$ is shown up to $t = 2.4$; it becomes infinite at $t = 160/63 \approx 2.54$. Figure 4.6-4 shows another $[-1,1]$-periodic FD4 solution, this one with $\theta = 0$ and random initial values of magnitude up to 0.15. The initial growth occurs where a local pattern similar to a section of (4.6-2) happens to be present. Quite often, the initial growth does not lead immediately to an explosion (as in this figure), but rather to an increased general noise level from which the growth process can repeat itself.

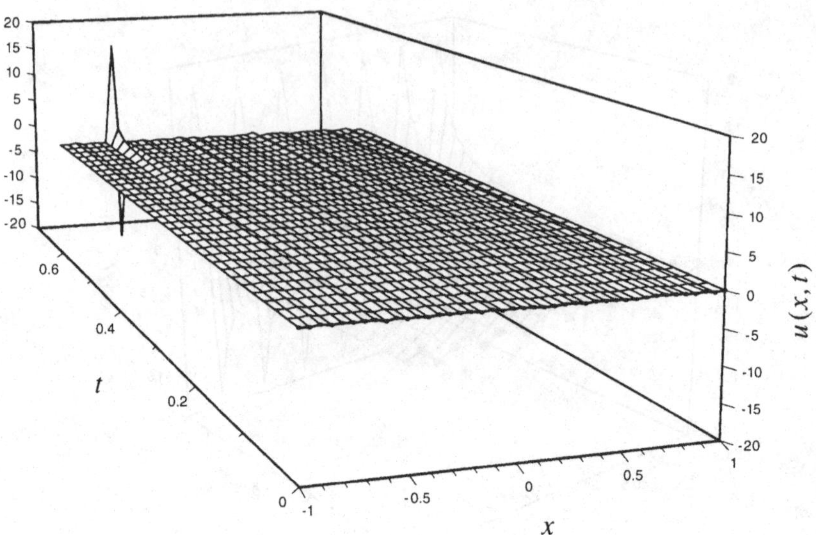

Figure 4.6-4. Periodic FD4 solution to $\partial u/\partial t + u(\partial u/\partial x) = 0$ with random initial values of magnitudes up to 0.15.

Nonlinear instabilities are difficult to anticipate, but in applications they arise infrequently. Corrective measures are normally taken only if "blow-ups" have occurred. Possible remedies include:

- weak viscous damping (needed only locally where a dangerous pattern can emerge – in this model problem, where the solution is near zero over a stretch of gridpoints or where a zero boundary condition is imposed); and
- use of an energy-conserving discretization (as exemplified by the choice $\theta = \frac{2}{3}$ for equation (4.6-1)). Arakawa (1966) describes a class of energy-conserving schemes for general convective equations (although their cost is a significant drawback).

The foregoing discussion focused on (4.6-1) because PS methods are especially prominent for equations dominated by convection (often in combination with small amounts of diffusion). However, nonlinear instabilities can arise also in other situations. The equation

$$u_t = u_{xx} + \sigma f(u) \tag{4.6-4}$$

serves as a simple model for many phenomena in chemistry and biology.

For example, (4.6-4) with $f(u) = e^u$ arises in combustion theory (with some solution blow-ups corresponding to ignition phenomena) and with $f(u) = u(1-u)$ in population genetics.

Equation (4.6-4) can be approximated by the FD2 scheme

$$\frac{u(x,t+k)-u(x,t)}{k}$$

$$=\theta\left[\frac{u(x+h,t+k)-2u(x,t+k)+u(x-h,t+k)}{h^2}+\sigma f(u(x,t+k))\right]$$

$$+(1-\theta)\left[\frac{u(x+h,t)-2u(x,t)+u(x-h,t)}{h^2}+\sigma f(u(x,t))\right], \qquad (4.6\text{-}5)$$

where θ is a free parameter (the choice $\theta=\frac{1}{2}$ corresponds to a standard Crank–Nicolson approximation). Stuart (1989) describes and compares different nonlinear instability mechanisms that are present in equations (4.6-4) and (4.6-5) for different values of σ and functions $f(u)$; in the case of (4.6-5), these also depend on θ and $\lambda=k/h^2$.

4.7. Gaussian quadrature formulas and PS methods

In the FD-based introduction of PS methods in Chapter 3, we noted that, as $N\to\infty$, it made sense to cluster the nodes roughly quadratically toward the ends of the interval. In many (nonperiodic) PS implementations, the nodes are chosen in a much more precise manner: exactly at the locations (knots) that arise in some Gaussian quadrature (GQ) formulas.

For example, Bernardi and Maday (1991) write, "We do think that the corner stone of collocation techniques is the choice of the collocation nodes", and "in spectral methods . . . these are always built from the nodes of a Gauss quadrature formula".

The main reasons for the focus on GQ formulas are that:

- the case of Chebyshev nodes (extrema) was the first one considered (possibly because it allowed the use of the FFTs), and this coincided with one of the classical GQ situations;
- stability and convergence analysis (by energy methods) can work out particularly conveniently (since it might be possible to replace sums by integrals; see Examples 4 and 5 in Appendix H);
- GQ knot locations are generally well suited as nodes for polynomial interpolation; and
- variational (e.g. Galerkin) formulations involve integrating products of polynomial test functions - GQ formulas lead to exact results even for quite high-degree polynomials.

The key question in this context is whether there are any major advantages to using GQ knot locations, and if so, whether these are sufficient to justify the effort in determining specialized GQ formulas (dependent on boundary conditions, etc.). The remainder of this section includes:

- a brief description of the GQ concept;
- a heuristic argument indicating why GQ knots are well suited as nodes for polynomial interpolation; and
- some numerical evidence suggesting that, after all, the selection of node locations is *not* a particularly critical issue.

We also illustrate in this section that the limit FD → PS for nonperiodic problems is *regular* in the sense that the accuracy (for smooth solutions) increases with the order of the FD scheme in a very steady manner, in contrast to the situation for quadrature if a GQ situation is encountered.

Gaussian quadrature formulas

Example 1. Determine a GQ formula (without any constraints on the knot positions) for $\int_{-1}^{1} f(x)\,dx$.

Independently of how the distinct knots x_i, $i=0,1,\dots,N$, are distributed, one can find weights w_i, $i=0,1,\dots,N$, such that the following integration formula becomes exact for all polynomials $r_N(x)$ of degree N:

$$\int_{-1}^{1} r_N(x)\,dx = \sum_{i=0}^{N} r_N(x_i)w_i. \tag{4.7-1}$$

Requiring this relation to hold for $r(x)=1,x,x^2,\dots,x^N$ gives rise to a nonsingular linear system that can be solved for $w_0,w_1,\dots,w_N$ (numerically, or in closed form; the coefficient matrix is of Van der Monde type).

GQ methods exploit the observation that certain choices of the knots x_i can make formulas such as (4.7-1) exact also for polynomials of much higher orders. Noting that any polynomial $p_{2N+1}(x)$ (of degree $2N+1$) can be written

$$p_{2N+1}(x)=s_N(x)P_{N+1}(x)+r_N(x)$$

where $P_{N+1}(x)$ is the Legendre polynomial of degree $N+1$, we have

$$\int_{-1}^{1} p_{2N+1}(x)\,dx$$

$$=\int_{-1}^{1} s_N(x)P_{N+1}(x)\,dx$$ $= 0$ since $s_N(x)$ can be written as a linear combination of $P_0(x),P_1(x),\dots,P_N(x)$ and $\int_{-1}^{1} P_i(x)P_{N+1}(x)\,dx=0$ for $i\le N$

$$+\int_{-1}^{1} r_N(x)\,dx$$

$$= \sum_{i=0}^{N} s_N(x_i)P_{N+1}(x_i)w_i$$ = 0 if we choose knots x_i as the zeros of $P_{N+1}(x)$ – all terms then vanish irrespectively of the values of $S_N(x_i)$ and of the weights w_i

$$+\sum_{i=0}^{N} r_N(x_i)w_i$$ where the weights w_i have been determined from the node locations x_i according to (4.7-1)

$$= \sum_{i=0}^{N} p_{2N+1}(x_i)w_i$$ using again $p_{2N+1}(x) = s_N(x)P_{N+1}(x)+r_N(x)$.

In other words, the quadrature formula (4.7-1) has, with the particular choice of knots x_i, become exact for all polynomials $p_{2N+1}(x)$ of degree $2N+1$.

The GQ formula in Example 1 corresponds to the top line in Table 4.7-1. The Radau and Lobatto variations can be derived following the same pattern as in Example 1. For instance, in the Chebyshev–Gauss–Radau case, we note that any polynomial $p_{2N}(x)$ can be written as

$$p_{2N}(x) = s_{N-1}(x)(T_N(x)+T_{N+1}(x))+r_N(x),$$

and we then include the weight function $1/\sqrt{1-x^2}$ in all the integrals.

Most textbooks in numerical analysis discuss GQ methods in some detail (including how to conveniently obtain closed-form expressions for the weights, which is not of particular relevance in the present PS context). In cases of weight functions $w(x)$ that are more general (than 1 and $1/\sqrt{1-x^2}$) and with more constraints on the form of the approximation, GQ formulas are most easily determined numerically. The algorithm by Golub and Kautsky (1983) provides the means for establishing the best positions of free knots, as well as the weights to be used at all knots, when the user specifies:

- the weight function $w(x)$;
- the location of some fixed knots (if any); and
- the amount of information to be used at the different knots (e.g., function value only, up to third derivative, etc. – the specifications may vary from knot to knot).

Suitability of GQ knots as PS nodes

Example 2. When interpolating a smooth function $f(x)$ with a polynomial $p_N(x)$ (of degree N), determine where the nodes x_i, $i = 0, 1, \ldots, N$,

Table 4.7-1. *Examples of Gaussian quadrature formulas*

Integral	Types of nodes	Form of approximation	Knot locations $x_k,\ k=0,\dots,N$
$\int_{-1}^{1} f(x)\,dx$	Interior only (Legendre-Gauss)		Zeros of P_{N+1}
	$f(-1)$ included (Legendre-Gauss-Radau)	$\sum_{k=0}^{N} w_k f(x_k)$	Zeros of $P_N + P_{N+1}$
	$f(\pm 1)$ included (Legendre-Gauss-Lobatto)		-1, zeros of P_N', $+1$
	$f(\pm 1)$ and $f'(\pm 1)$ included	$\check{w}_0 f'(x_0) + \check{w}_N f'(x_N) + \sum_{k=0}^{N} w_k f(x_k)$	-1, zeros of P_{N+1}'', $+1$
$\int_{-1}^{1} \frac{f(x)}{\sqrt{1-x^2}}\,dx$	Interior only (Chebyshev-Gauss)		$\cos \frac{(2k+1)\pi}{2N+2}$
	$f(-1)$ included (Chebyshev-Gauss-Radau)	$\sum_{k=0}^{N} w_k f(x_k)$	$\cos \frac{2\pi k}{2N+1}$
	$f(\pm 1)$ included (Chebyshev-Gauss-Lobatto)		$\cos \frac{\pi k}{N}$
	$f(\pm 1)$ and $f'(\pm 1)$ included	$\check{w}_0 f'(x_0) + \check{w}_N f'(x_N) + \sum_{k=0}^{N} w_k f(x_k)$	-1, zeros of T_{N+1}'', $+1$

Weights w_k, $k=0,\dots,N$	Exact for polynomials up to degree:
$w_k = \dfrac{2}{(1-x_k^2)[P'_{N+1}(x_k)]^2}$	$2N+1$
$w_0 = \dfrac{2}{(N+1)^2}$ $(k=0)$ $w_k = \dfrac{1}{(N+1)^2}\dfrac{1-x_k}{[P_N(x_k)]^2}$ $(k\geq 1)$	$2N$
$w_k = \dfrac{1}{N(N+1)}\dfrac{2}{[P_N(x_k)]^2}$	$2N-1$
$\check{w}_0 = -\check{w}_N = \dfrac{8}{N(N+1)(N+2)(N+3)}$ $(k=0,N)$ $w_0 = w_N = \dfrac{8(2N^2+6N+1)}{3N(N+1)(N+2)(N+3)}$ $(k=0,N)$ $w_k = \dfrac{2N(N+1)(N+2)}{(N+3)}\dfrac{1}{(1-x_k^2)[P''_N(x_k)]^2}$ $(1\leq k\leq N-1)$	$2N+1$
$w_k = \dfrac{\pi}{N+1}$	$2N+1$
$w_0 = \dfrac{\pi}{2N+1}$ $(k=0)$ $w_k = \dfrac{2\pi}{2N+2}$ $(k\geq 1)$	$2N$
$w_0 = w_N = \dfrac{\pi}{2N}$ $(k=0,N)$ $w_k = \dfrac{\pi}{N}$ $(1\leq k\leq N-1)$	$2N-1$
$\check{w}_0 = -\check{w}_N = \dfrac{3\pi}{4N(N+1)(N+2)}$ $(k=0,N)$ $w_0 = w_N = \dfrac{3\pi(3N^2+6N+1)}{10N(N+1)(N+2)}$ $(k=0,N)$ $w_k = \dfrac{\pi N^3(N+1)}{N+2}\dfrac{1}{(1-x_k^2)[T''_N(x_k)]^2}$ $(1\leq k\leq N-1)$	$2N+1$

should be placed in order to minimize the L^2 norm $\|f(x)-p_N(x)\|_2^2 = \int_{-1}^{1}(f(x)-p_N(x))^2\,dx$.

Heuristic solution: If $f(x)$ is expanded in Legendre polynomials $f(x) = \sum_{k=0}^{\infty} a_k P_k(x)$, the best (in L^2 norm) polynomial approximation is the truncated expansion

$$f_N(x) = \sum_{k=0}^{N} a_k P_k(x),$$

since this is orthogonal to the remainder

$$r_N(x) = \sum_{k=N+1}^{\infty} a_k P_k(x).$$

If we somehow knew $f_N(x)$ and could interpolate it to order N (instead of interpolating $f(x)$), we would exactly recover $f_N(x)$ independently of how the nodes x_i were selected. With only $f(x)$ available, the best general strategy would be to interpolate where the remainder is the smallest:

$$\begin{aligned} f(x)-f_N(x) &= r_N(x) && \text{(remainder)} \\ &= a_{N+1}P_{N+1}(x) && \text{(dominant term)} \\ &\quad + a_{N+2}P_{N+2}(x) && \text{(smaller terms)} \\ &\quad + \cdots. \end{aligned}$$

The dominant term vanishes if we choose the x_i as the zeros of $P_{N+1}(x)$.

For large deviations from these Legendre–Gauss positions, a deteriorating Lebesgue constant may add to the errors. For small deviations, the Lebesgue constant remains small and its effect is minor.

The interpolation error in Example 2 is minimized by the same node selection as for Example 1. However, there is a major difference between these two cases regarding the importance of making precisely this node selection:

Quadrature – the result becomes exact for much higher-order polynomials (degree $2N+1$ vs. N).

Interpolation – the gain is only one term in the error expansion (corresponding to a gain of only one degree for polynomials).

Sensitivity to node locations – numerical comparison

To illustrate the sensitivities to knot (node) locations, we consider the following two problems.

1. *Eigenvalue problem:* $u_{xx} = \lambda u$, $u(\pm 1) = 0$. The analytic eigenvalues are $\lambda_k = -(k\pi/2)^2$, $k = 1, 2, 3, \ldots$ (cf. Example 3 in Section 4.4).
2. *Numerical quadrature:* $\int_{-1}^{1}(\cos(k\pi x)/\sqrt{1-x^2})\,dx = \pi J_0(k\pi)$ for all k, where $J_0(x)$ is the Bessel function of order zero. These integrals are considered here for $k = 1, 2, 3, \ldots$.

Example 3. Compare the PS accuracies for solutions to Problems 1 and 2 when the grid-density parameter γ (see Section 3.3) is varied around 0.5.

Problem 1: We form the PS DM for u_{xx} and disregard its first and last row and column. Figure 4.7-1(a) shows how many correct decimal places are obtained for different eigenvalues k and different γ. As expected from

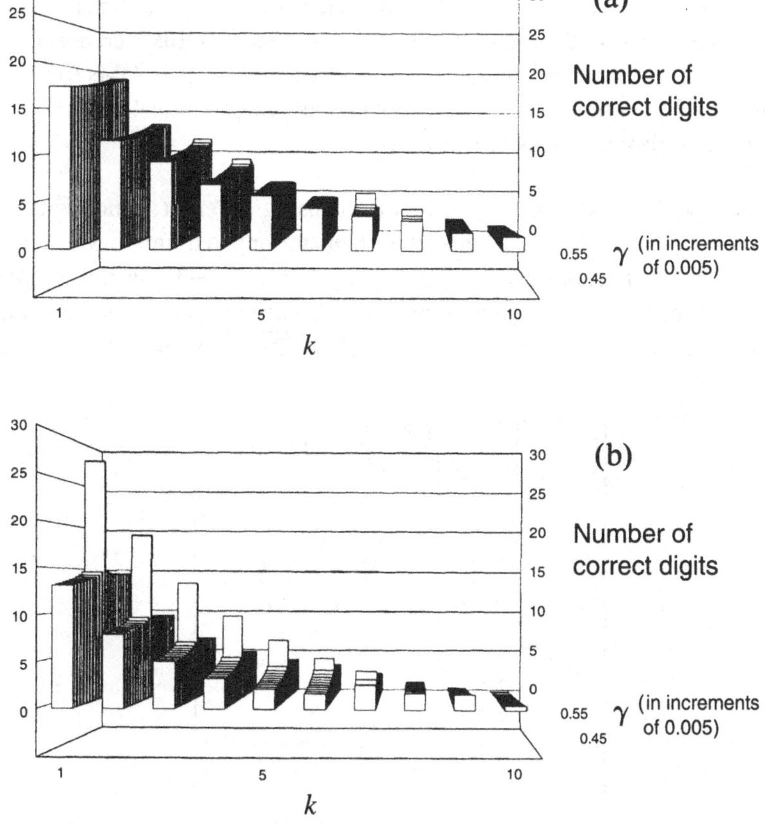

Figure 4.7-1. Number of correct digits for different values of k and γ. **(a)** Eigenvalue problem, $N = 16$. **(b)** Numerical quadrature, $N = 16$.

Figure 4.4-2, very high accuracy is obtained consistently for the lowest eigenvalues.

Problem 2: To integrate $\int_{-1}^{1}(f(x)/\sqrt{1-x^2})\,dx$, we approximate $f(x)$ by polynomial interpolation through all the node points. The resulting integral is then evaluated in closed form. Figure 4.7-1(b) shows the number of correct decimal places obtained for different k and γ – in this case, featuring an extreme sensitivity to deviations in γ away from 0.5.

Example 4. Compare the accuracies for solutions to Problems 1 and 2 when approximated with FD-type methods of increasing accuracies on a standard Chebyshev grid.

Problem 1: We approximate u_{xx} in standard FD fashion at each interior node point using stencils of widths 3 (the minimal possible for a second derivative), $5, 7, \dots, N+1$ (the PS limit). Near the ends, the stencils need to be partly one-sided – increasingly so as the stencil widths increase. Figure 4.7-2 shows schematically the nonzero structure of these DMs for u_{xx}. Figure 4.7-3(a) shows how the accuracy increases systematically as higher-order FD approximations are employed.

Problem 2: The integral is approximated using different stencil widths $m = 2, 4, 6, \dots, N+1$. For each value of m, we approximate first the integral over each subinterval $[x_i, x_{i+1}]$, $i = 0, 1, \dots, N-1$, by interpolating $f(x) = \cos(k\pi x)$ by a polynomial $p_{m-1}(x)$ at the gridpoints $\{x_j, x_{j+1}, \dots, x_{j+m-1}\}$. We then evaluate $\int_{x_i}^{x_{i+1}} p_{m-1}(x)/\sqrt{1-x^2}\,dx$ analytically. The

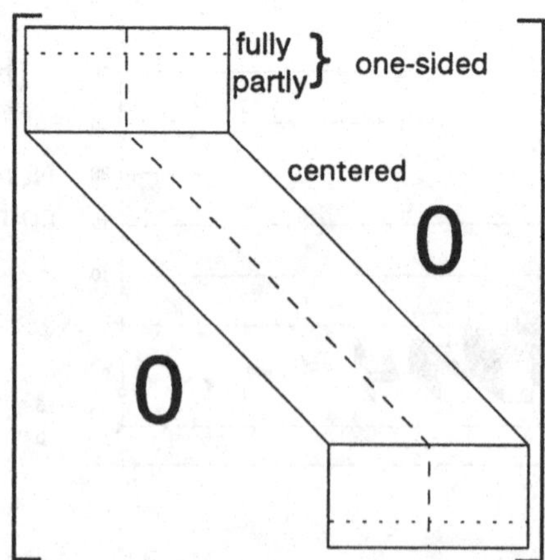

Figure 4.7-2. Structure of DM for u_{xx} when using an FD method.

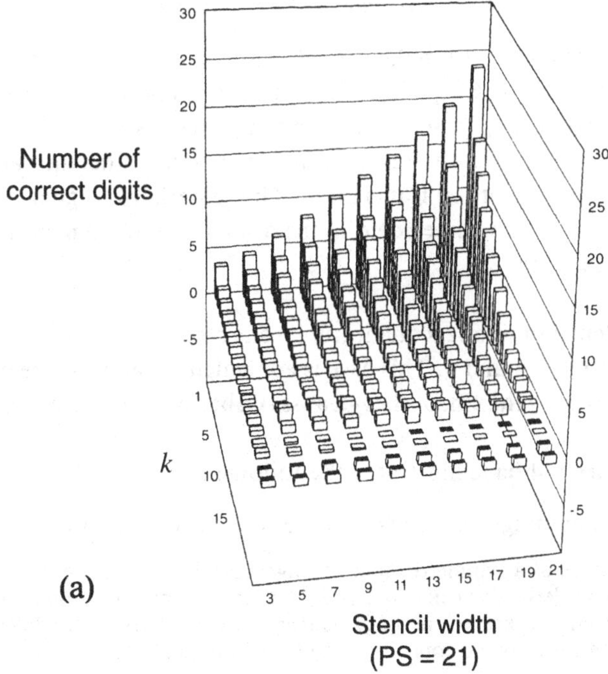

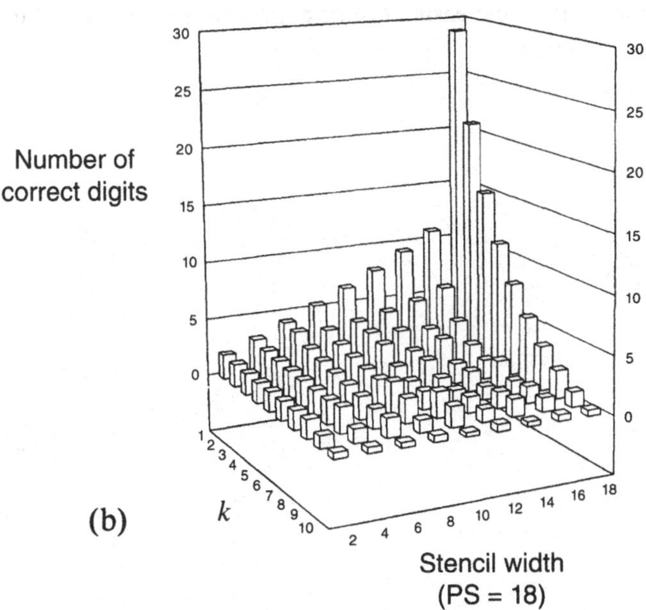

Figure 4.7-3. Number of correct digits for different values of k and different stencil widths. **(a)** Eigenvalue problem, $N = 20$. **(b)** Numerical quadrature, $N = 17$.

value for j is chosen to make the interval $[x_j, x_{j+m-1}]$ as symmetrically located, relative to $[x_i, x_{i+1}]$, as the boundaries allow. For each value of m, the contributions from the different subintervals $[x_i, x_{i+1}]$ are added. Figure 4.7-3(b) shows the resulting accuracies for different values of k when the stencil width m is increased. The remarkably high accuracy for the GQ formula is again evident in the dramatic improvement from $m = 16$ to $m = N+1 = 18$.

For Problem 2 (numerical quadrature), one single case stands out in both of the Examples 3 and 4. This case is equivalent to an instance of Gaussian quadrature – the Chebyshev–Gauss–Lobatto formula shown in Table 4.7-1.

Examples 1–4 have illustrated the following.

(1) Gaussian integration formulas are spectacularly accurate.

In the case of a symmetric node arrangement, a nonperiodic PS approximation for a derivative can *at one point* be exact for all polynomials *one degree above* what is expected in the general case. In other words, for differentiation there seems to be no counterpart to Gaussian quadrature.

(2) PS methods for derivatives are nowhere nearly as sensitive to node perturbations as are Gaussian quadrature formulas.

(3) FD methods of increasing orders approach the PS method in a very regular manner.

5

PS variations and enhancements

Up to this point, we have described basic PS implementations. However, many variations are possible, offering advantages in different respects. In this chapter, we discuss a few of these variations.

5.1. Use of additional information from the governing equations

This idea (like most others) is best described through the use of examples. It requires that the problem be manipulated analytically (e.g., by repeated differentiation) to provide more information than is immediately available from its original formulation.

Example 1. Exploit additional derivative information at the boundaries when solving the eigenvalue problem $u_{xx} = \lambda u$, $u(\pm 1) = 0$.

Since $u(\pm 1) = 0$, clearly also $u''(\pm 1) = 0$ and $u''''(\pm 1) = 0$ (we label these as "extra" boundary conditions - for this example, we ignore that this pattern continues indefinitely and that u becomes periodic). The boundary information on u'' and u'''' can be exploited in different ways.

A. Reduce the largest spurious EVs (cf. Figure 4.4-2). To each extra boundary condition (such as $u''(-1) = 0$) corresponds a one-sided difference stencil. From each row of the DM, like those shown in Figures 4.3-1(c) and (d), we can subtract any multiples of these stencils without compromising the spectral accuracy. The multiples can be chosen to minimize the sum of the squares of the elements of the resulting DM. As shown in Fornberg (1990b), this procedure significantly improves the conditioning of the DMs:

- Using only $u''(\pm 1) = 0$ and $u''''(\pm 1) = 0$ reduces the largest matrix elements of the DMs shown in Figures 4.3-1(c) and (d) (edge elements not included) to about 500 - down from (resp.) about 500,000 and 7,000.

- Each time an extra boundary condition is applied to a Chebyshev-type approximation, the largest (remaining) spurious eigenvalue is changed to *exactly* zero.

Extremely large and zero EVs are both wrong, but the latter is much easier to handle. One convenient way to proceed in time-dependent problems is to apply a little bit less than the described changes (multiples) – say, 90% or 99% of these. The spurious EVs are then not reduced to exactly zero, but remain sufficiently negative to dampen out rapidly. The modified scheme may then be well suited for explicit time stepping; see the comments near the end of the Example in Section 7.2.

B. Increase the accuracy of the computed EVs. Given k extra relations (e.g., information at the boundaries), we can proceed as follows.

(1) Introduce k new fictitious gridpoints (anywhere – inside or outside the domain).
(2) Find the $N+1$ stencils for u_{xx} that are accurate at the original gridpoint locations $x_0, x_1, \ldots, x_N$ respectively, but that extend also over the fictitious points.
(3) Find the k stencils that express the k extra pieces of information (again extending over all gridpoints, original and fictitious).
(4) Add/subtract multiples of these k stencils to/from the ones calculated in step (2), so that the weights at all the fictitious points are eliminated.

The resulting stencils for u_{xx} become exact for all polynomials of degree $N+k$ that take prescribed values at $x_0, x_1, \ldots, x_N$ *and* satisfy the k extra relations. They are therefore more accurate than straightforward PS approximations that also extend over $x_0, x_1, \ldots, x_N$ (which would be exact only for polynomials up to degree N).

If the "side conditions" are not available explicitly (e.g. $f'''(-1) = \alpha$, as opposed to $f'''(-1) = 7$), the four steps just listed can instead be used to rapidly generate the weights $\{c_j\}$, $j = 0, \ldots, N$, and $\{d_j\}$, $j = 0, \ldots, M$, in FD (or PS) formulas such as

$$f^{(k)}(\xi) = \sum_{j=0}^{N} c_j f(x_j) + \sum_{j=0}^{M} d_j f^{(k_j)}(\xi_j),$$

where $\{k_j\}$ is any sequence of positive integers and $\{\xi_j\}$ any set of x coordinates, $j = 0, \ldots, M$. This requires the use of $M+1$ fictitious points (which need to be distinct from the points $\{x_j\}$, $j = 0, \ldots, N$).

In the special case when the derivative information is available at the same locations as the function values (Hermite interpolation), the weights in such formulas as

$$f^{(k)}(\xi) \approx \sum_{j=0}^{i} c_{i,j}^{k} f(x_j) + \sum_{j=0}^{i} d_{i,j}^{k} f'(x_j), \quad k=0,1,\ldots,m, \; i=\left[\frac{k}{2}\right], \left[\frac{k}{2}\right]+1, \ldots, n,$$

(cf. equation (3.1-1)) are best found from a generalization of the algorithm in Section 3.1 (Fornberg 1996).

The technique of introducing fictitious points outside a boundary and then eliminating them by means of boundary conditions is often used with FD methods. The PS case is remarkable in that the location of these points turns out to have no influence at all on the final result (apart from rounding errors). In the PS case, this technique offers a very convenient way of generating stencils that satisfy side conditions, without the need to develop any additional analytical devices.

Both methods in Example 1 can be applied to many time-dependent problems. In the case of the heat equation $u_t = u_{xx}$, $u(\pm 1) = 0$, we again obtain $u''(\pm 1) = 0$, $u''''(\pm 1) = 0, \ldots$.

The fictitious-point method just described offers a very convenient way to approximate high-order linear two-point boundary value problems, as shown in the following example.

Example 2. Solve $u''''(x) = f(x)$, where $f(x)$, $u(\pm 1)$, and $u'(\pm 1)$ are given in such a way that $u(x) = 1 + \sin(2\pi x)$ becomes a solution.

The following steps give spectral solutions consistent with the boundary conditions (BCs).

(1) Add as many fictitious points as needed in order to obtain an equal number of unknowns as equations (i.e., add two points if we choose to approximate the governing equation at interior points only, four if we approximate it also at $x = \pm 1$).
(2) Use *all* gridpoints when approximating the governing equation (at the regular gridpoints only) and the boundary conditions.
(3) Solve the resulting linear system.
(4) Ignore the values at fictitious points.

As before, the locations of the fictitious points affect only rounding errors. For example, they can be placed just outside the boundaries.

The two curves in Figure 5.1-1 show $\log_{10}$ of the largest pointwise error in this example when two and four fictitious points were used, respectively. The markers along the right edge show the errors when the points were not distributed according to the γ parameter (cf. Section 3.3), but according

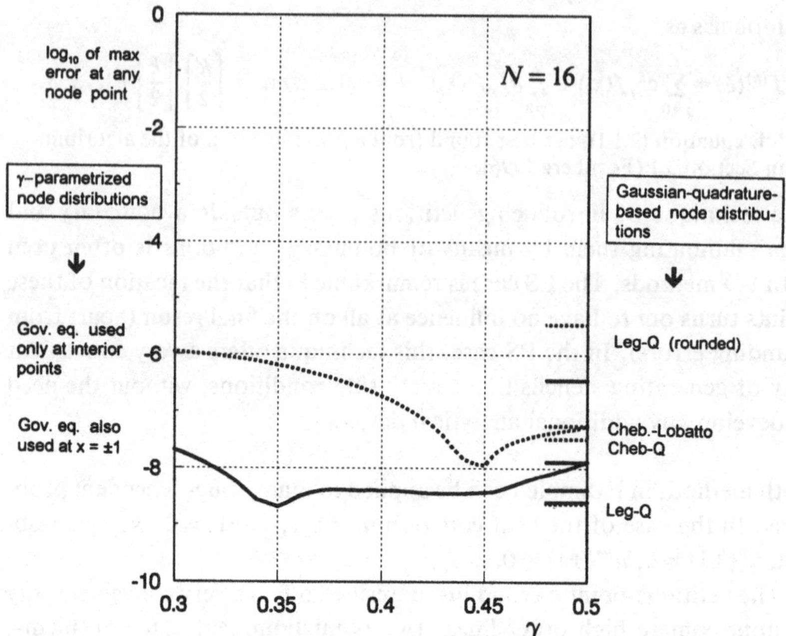

Figure 5.1-1. Accuracies of different PS solution techniques applied to $u''''(x) = f(x)$, where $f(x)$, $u(\pm 1)$, and $u'(\pm 1)$ were selected so that $u(x) = 1 + \sin(2\pi x)$. See text.

to different Gaussian quadrature rules (cf. Table 4.7-1). The figure supports our observation in Section 4.7 that placing the nodes according to Gaussian quadrature knots can be expected to work very well, but is not critically important for the performance of PS methods.

Two different approximation strategies were employed at $x = \pm 1$. Accuracy data for these are marked as follows. *Dashed* - BCs, but not governing equation, enforced at $x = \pm 1$. *Solid* - BCs and governing equation enforced at $x = \pm 1$.

For each of these two boundary procedures, one curve and four edge markers show the accuracy that is obtained when the nodes are distributed as follows. *Curves* - Nodes distributed according to different γ values in $[0.3, 0.5]$. *Right edge markers* - Nodes distributed as follows (from top to bottom).

Leg-Q (rounded): same as Leg-Q, but with each node coordinate rounded to two decimal places; i.e., a 0.005-level noise in the node positions.

Cheb.-Lobatto: standard Chebyshev node distribution; identical to the case $\gamma = 0.5$.

Cheb-Q: nodes at [-1, zeros of T''_{N+1}, $+1$].

Leg-Q: nodes at [-1, zeros of P''_{N+1}, $+1$].

The cases marked "Leg-Q (rounded)" (top two markers along the right edge) show that noise in the node positions significantly degrades the resulting PS accuracy. Relatively smooth node distributions appear to be preferable to more irregular ones.

Another boundary case for which large improvements can be achieved concerns the origin in polar coordinates. This is discussed further in Sections 6.2 and 6.3.

Example 3. Exploit symmetry at the boundary $r=0$ when solving the eigenvalue problem

$$u''+\frac{1}{r}u'-\frac{n^2}{r^2}u=-\lambda u, \quad n=0,1,\dots, \; u(0) \text{ bounded}, \; u(1)=0.$$

This equation arises from the separation of variables in Poisson's equation on the unit circle (which suggests that u can be expected to continue in a very regular manner across $r=0$). The exact eigenvalues $\lambda_{n,k}$, $k=1,2,\dots,$ satisfy $J_n(\sqrt{\lambda_{n,k}})=0$ where $J_n(x)$ is a Bessel function of order n.

We compare two approximation methods.

1. Note that

$$u'(0)=0 \quad \text{if } n=0,$$
$$u(0)=0 \quad \text{if } n\neq 0,$$

and use straightforward Chebyshev approximation on $[0,1]$. Gridpoints are located at $r_k=(1-\cos(k\pi/N))/2$, $k=0,1,\dots,N$ (i.e., the grid is clustered at $r=0$ as well as at $r=1$).

2. Note that

$$u(r) \text{ is even} \quad \text{if } n \text{ is even},$$
$$u(r) \text{ is odd} \quad \text{if } n \text{ is odd}.$$

Consider FD stencils extending over $[-1,1]$, but use symmetry (or antisymmetry) to reduce actual calculations to within $[0,1]$. Gridpoints are located at $r_k=\sin(k\pi/2N)$, $k=0,1,\dots,N$ (i.e., the grid is clustered at $r=1$ only, not at $r=0$).

Figure 5.1-2 compares the accuracies of the numerical EVs obtained through these two methods for $n=7$ and $k=1$ ($\lambda_{7,1}\approx 122.9$, cf. Gottlieb and Orszag 1977, pp. 152–3). The values for method 1 are taken from Huang and Sloan (1993a). For comparison, values for equi-spaced FD methods of orders 2 and 4 are also shown.

Further variations of boundary implementations are discussed by Canuto and Quarteroni (1987) and by Funaro and Gottlieb (1988).

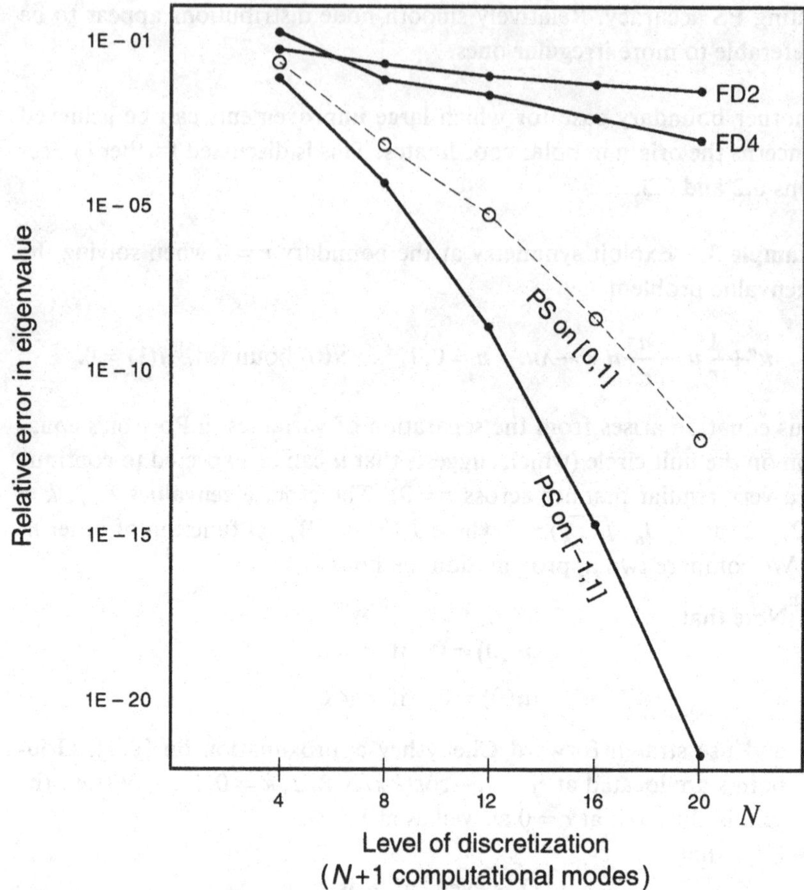

Figure 5.1-2. Errors in eigenvalue $\lambda_{7,1}$ for Bessel's equation when approximated using four different methods: second- and fourth-order FD methods, and PS approximations with and without treating the origin as a domain boundary.

The foregoing examples illustrate that:

- when using PS methods, one should always consider whether the FD viewpoint can offer any advantages (in accuracy, simplicity, flexibility, etc.); and
- the fundamental reason for clustering gridpoints at the ends of an interval is to compensate for the large error terms in one-sided approximations. Exploiting extra information at the boundaries both increases accuracy and decreases the need for grid clustering.

5.2. Use of different PS approximations for different terms in an equation

When a single variable appears more than once in an equation, it is normally approximated in a similar way at each instance. However, Huang and Sloan (1993b, 1994) note two situations when it is better to use different types of PS approximations.

Example 1. Solve the singular perturbation problem $\epsilon u'' + u' = 1$, $x \in [-1,1]$, $0 < \epsilon \ll 1$.

Straightforward centered FD2 approximations for both u' and u'' give an oscillatory solution with $O(1)$ errors across $[-1,1]$ for any N when $\epsilon < 1/N$. Approximating u_x by the one-sided FD stencil $[u(x+h) - u(x)]/h$ reduces the errors to $O(1/N)$. When using Chebyshev PS approximations, we can similarly approximate u' with stencils based on all gridpoints *except* the one at $x = -1$. In the limit of $\epsilon \to 0$, this gives spectral accuracy across $[-1,1]$ (rather than $O(1)$ errors).

For another approach to solving this problem with a PS method, see Eisen and Heinrichs (1992).

Example 2. Solve the eigenvalue problem $u'''' + 4u''' = \lambda u''$, $x \in [-1,1]$, $u(\pm 1) = u'(\pm 1) = 0$.

Problems similar to this arise, for example, in linearized stability analysis in fluid mechanics. In this case, spurious EVs are those appearing incorrectly in the right half-plane, suggesting physical instabilities that do not exist. If we approximate all derivatives of u on a Chebyshev grid, incorporating $u'(\pm 1) = 0$ as in Example 1B of Section 5.1, we will obtain spurious EVs. However, we can overcome this difficulty by ignoring $u'(\pm 1) = 0$ when approximating u''.

In both of these examples, variable coefficients would have added no complications, as is usually the case for PS methods (in sharp contrast to spectral Galerkin or tau methods).

5.3. Staggered grids

When using an FD method, it is customary to compute values for each unknown at each gridpoint. However, staggered grids often help both stability and accuracy in applications such as fluid dynamics and elasticity.

2-D ELASTIC WAVE EQUATION

$$\begin{cases} \rho u_t = f_x + g_y \\ \rho v_t = g_x + h_y \\ f_t = (\lambda + 2\mu) u_x + \lambda v_y \\ g_t = \mu v_x + \mu u_y \\ h_t = \lambda u_x + (\lambda + 2\mu) v_y \end{cases}$$

u, v velocities in x and y directions

f, g, h stress components

ρ, λ, μ given functions of x and y, material constants

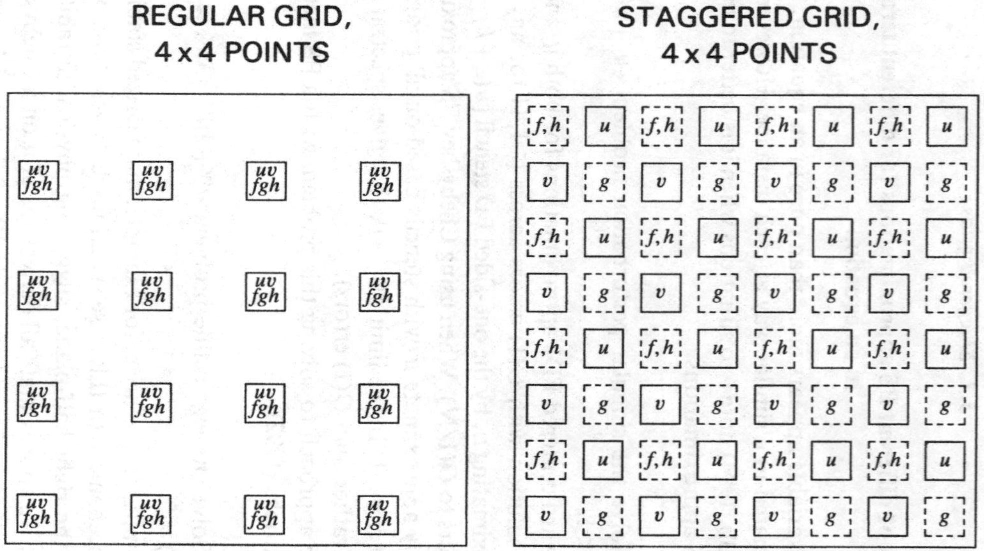

REGULAR GRID, 4 x 4 POINTS

uv fgh	uv fgh	uv fgh	uv fgh
uv fgh	uv fgh	uv fgh	uv fgh
uv fgh	uv fgh	uv fgh	uv fgh
uv fgh	uv fgh	uv fgh	uv fgh

STAGGERED GRID, 4 x 4 POINTS

f,h	u	f,h	u	f,h	u	f,h	u
v	g	v	g	v	g	v	g
f,h	u	f,h	u	f,h	u	f,h	u
v	g	v	g	v	g	v	g
f,h	u	f,h	u	f,h	u	f,h	u
v	g	v	g	v	g	v	g
f,h	u	f,h	u	f,h	u	f,h	u
v	g	v	g	v	g	v	g

Figure 5.3-1. Example of staggered grid arrangement.

Figure 5.3-1 illustrates this idea for equations relevant to seismic exploration.

Another example is provided by the two Maxwell curl equations

$$\epsilon \frac{\partial \vec{E}}{\partial t} = \nabla \times \vec{H} - \sigma \vec{E},$$

$$\mu \frac{\partial \vec{H}}{\partial t} = -\nabla \times \vec{E},$$

where the unknowns $H_{1,2,3}$ and $E_{1,2,3}$ can be laid out in a cubic lattice at the midpoints of the faces and the edges respectively. Perfect staggering is again achieved for each equation and each unknown. As in the 2-D elastic example, spatial staggering can be combined with time staggering when using, for example, leapfrog to advance in turn the H and E components.

The gain in accuracy by spatial staggering is well known for FD2 approximations of the first derivative (in which case staggering effectively halves the space step). The benefit in accuracy turns out to increase with the order of the approximation, as the following table (for the first derivative) shows.

Grid type	Approximation	Leading error term	Ratio of error terms
Second-order accuracy			
Regular	$f'(x) = [-\frac{1}{2}f(x-h)+\frac{1}{2}f(x+h)]/h$	$+\frac{h^2}{6}f'''(x)$	$\frac{1}{4} = 0.25$
Staggered	$f'(x) = [-f(x-\frac{h}{2})+f(x+\frac{h}{2})]/h$	$+\frac{h^2}{24}f'''(x)$	
Fourth-order accuracy			
Regular	$f'(x) = [\frac{1}{12}f(x-2h)-\frac{2}{3}f(x-h)$ $+\frac{2}{3}f(x+h)-\frac{1}{12}f(x+2h)]/h$	$-\frac{h^4}{30}f^{v}(x)$	$\frac{9}{64} \approx 0.141$
Staggered	$f'(x) = [\frac{1}{24}f(x-\frac{3h}{2})-\frac{9}{8}f(x-\frac{h}{2})$ $+\frac{9}{8}f(x+\frac{h}{2})-\frac{1}{24}f(x+\frac{3h}{2})]/h$	$-\frac{3h^4}{640}f^{v}(x)$	

For approximations of order p (even), the ratio of error terms becomes $(p!)^2/[2^p((p/2)!)^2]^2 \approx 2/\pi p$. Since the periodic PS method can be viewed as the limit of $p \to \infty$, this suggests that the idea of staggering should be advantageous in that case also.

Another suggestive argument follows from comparing the weights in the stencils. Figure 5.3-2(a) shows the same information as in the right half of Figure 3.2-1 – the magnitudes of the weights for increasingly accurate approximations to the first derivative. In the limit, these become $(-1)^j/j$, $j = 1, 2, \ldots$. For the staggered approximations, the limit is much more local in nature: $(-1)^{j+1/2}/\pi j^2$, $j = \frac{1}{2}, \frac{3}{2}, \frac{5}{2}, \ldots$. Since the derivative is a local property of a function, a more compact approximation makes

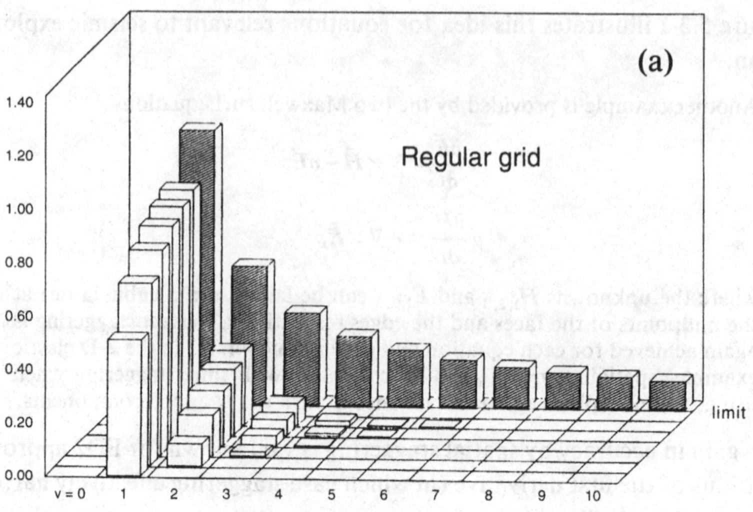

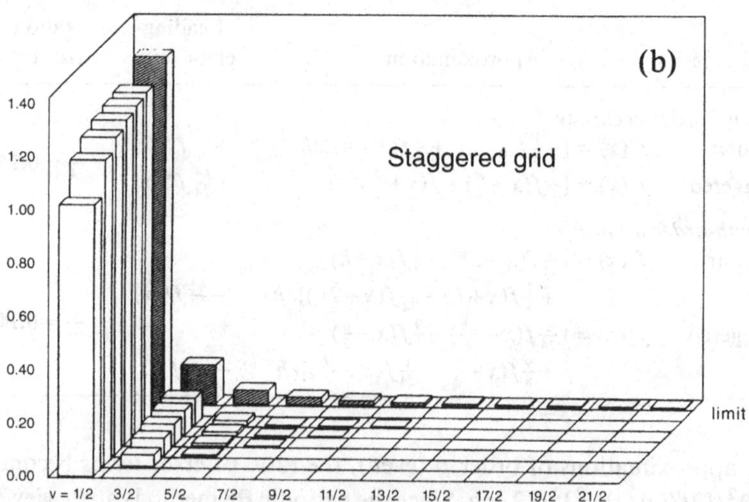

Figure 5.3-2. Magnitudes of weights for increasingly accurate approximations to the first derivative on (a) regular and (b) staggered grids (right halves of stencils displayed).

more sense than one relying on extensive cancellation of distant contributions.

In this case of equi-spaced grids, staggering proves advantageous for odd derivatives (first, third, etc.), whereas regular grids are better for even derivatives (see Fornberg 1990a). For nonperiodic problems, staggering can be achieved by using grids based on Chebyshev extrema (as usual) and Chebyshev zeros (in Gaussian quadrature nomenclature, Lobatto and regular node locations, resp.; cf. Section 4.7).

The fundamental idea behind staggering is that the locations where function values are available (gridpoints) need not coincide with the locations at which we approximate the governing equations. In Section 5.1, both Example 1B and Example 2 relate to this very general concept. In a nonperiodic PS method, additional information at any number of locations can be incorporated by means of adding a matching number of fictitious gridpoints.

5.4. Preconditioning

The large spurious eigenvalues in many PS DMs can make explicit time stepping methods very costly (forcing the use of extremely small time steps). Implicit methods often have unbounded stability domains, but they require the solution of a full linear system at every time step.

For equations in one dimension, Gaussian elimination tends to be both affordable and trouble-free. However, the cost increases rapidly with the number of unknowns, making iterative methods preferable in two or more dimensions. Preconditioning exploits knowledge about the underlying problem to produce linear systems that can be solved effectively.

Example. Precondition for the Chebyshev DM solving the two-point boundary value problem $u''(x) = f(x)$, where $u(\pm 1)$ and $f(x)$ are given.

Chebyshev PS discretization (viewed as an FD method) gives rise to a linear system $Cu = f$, where C is the Chebyshev DM for d^2/dx^2 (as shown in Figure 4.3-1(d) but with the edge rows and columns removed). C is neither symmetric nor diagonally dominant, so standard iterative techniques will not converge.

Let F be the second-order FD DM based on the same Chebyshev grid (with elements obtained using the algorithm in Section 3.1). Since F is tri-diagonal, it is easily inverted (in higher dimensions, one can use alternating direction arrangements of tri-diagonal matrices). The system to be solved can be written as $[F^{-1}C]u = g$, where $g = F^{-1}f$. The matrix $F^{-1}C$ is illustrated in Figure 5.4-1. In sharp contrast to the matrix C, the matrix $F^{-1}C$ is strongly diagonally dominant. The new linear system is very well suited for iterative solution. Any standard iterative technique can be used to rapidly obtain the (spectrally accurate) vector u.

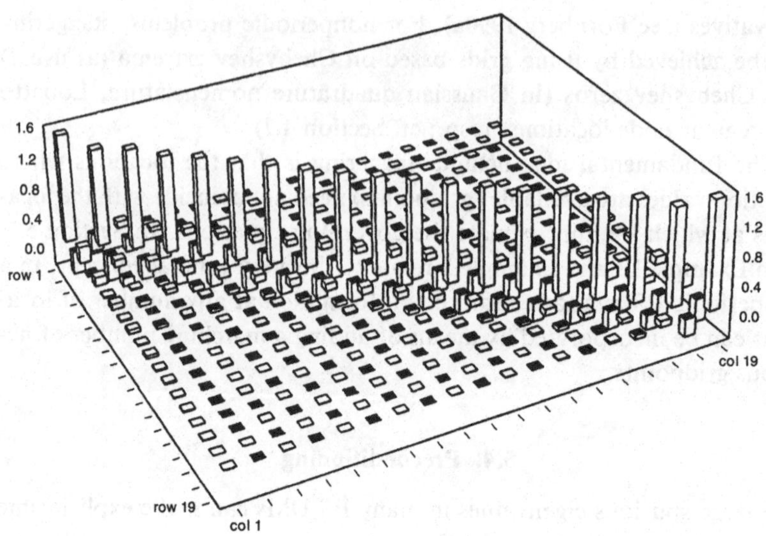

Figure 5.4-1. Display of the matrix $F^{-1}C$ resulting from FD2 preconditioning of the Chebyshev DM C for d^2/dx^2 (with $N=20$).

For matrices of this form (near-symmetric and diagonally dominant), convergence of some methods improve if the ratio of largest to smallest eigenvalues is lowered. For $F^{-1}C$, $\lambda_{max}/\lambda_{min} \to \pi^2/4$ as $N \to \infty$ (Haldenwang et al. 1984). Using higher-order FD methods, this ratio can be lower still (Phillips, Zang, and Hussaini 1986), but savings may be offset by a higher cost of applying F^{-1}. In contrast, we note that for the (nonsymmetric, non-diagonally dominant) matrix C, $\lambda_{max}/\lambda_{min}$ grows like $O(N^4)$.

The idea of preconditioning with second-order FD methods is outlined in Orszag (1980). For odd derivatives, FD preconditioning normally benefits from the use of staggered grids. This is discussed for Chebyshev methods by Hussaini and Zang (1984) and Funaro (1987). Mulholland and Sloan (1992) consider FD preconditioners for staggered approximations to $\partial^3/\partial x^3$ (applicable e.g. to the Korteweg–de Vries equation).

Finite element-based preconditioners have been discussed by Canuto and Quarteroni (1985), Deville and Mund (1985), and Canuto and Pietra (1987). General references on preconditioning include Canuto et al. (1988) and Boyd (1989).

5.5. Change of independent variable

All approximations to derivatives that we have considered so far (for nonperiodic problems) have been based on differentiating interpolating polynomials. Difficulties at boundaries have been linked to large weights in one-sided stencils, a consequence of the rapid growth of high-degree polynomials at increasing distances from the origin.

A change of independent variable can bypass some less desirable properties of polynomials while keeping their advantages (spectral accuracy for PS methods, ease of computation, etc.). We can replace $x \in [-1,1]$ by $y = y(x) \in [-1,1]$ and then use, say, the Chebyshev–PS method in the y-coordinate system. In the original x-coordinate system, this corresponds to a non-polynomial-based PS method over non-Chebyshev-distributed gridpoints. This approach is fundamentally different from applying a polynomial-based PS method over the x-coordinate gridpoints. Polynomial basis functions do not gain in their ability to resolve details if gridpoints are clustered in some interior area (the main effect will be a much worsened conditioning). We shall now describe three situations in which a change of variable can be particularly advantageous.

Increased resolution at interior or boundary locations

Bayliss et al. (1989) and Bayliss and Turkel (1992) propose a change of variables to enhance the local resolution in problems with steep moving fronts (arising e.g. in compressible fluid dynamics, combustion, etc.). Tang and Trummer (1994) use changes of variable to resolve boundary layers in singular perturbation problems.

Improving the condition number of DMs

Kosloff and Tal-Ezer (1993) propose to first change the independent variable x into y through $x = \arcsin(\alpha y)/\arcsin(\alpha)$ (both x and $y \in [-1,1]$; the parameter $\alpha \in [0,1]$). In the governing equations, $\partial/\partial x$ must then be replaced by $(\arcsin(\alpha)/\alpha)\sqrt{1-(\alpha y)^2}(\partial/\partial y)$ (and similarly for higher derivatives). Applying a standard Chebyshev–PS method in the y variable will correspond, in the x variable, to working with nonpolynomial basis functions.

In the limit of $\alpha \to 0$, y becomes equal to x, and we have the regular Chebyshev–PS method. As $\alpha \to 1$, the x-coordinate grid approaches uniform spacing. Close to this limit, the Chebyshev polynomials (in the y variable) have, in the x variable, become stretched to resemble trigonometric functions. This reduces the spurious EVs; in Example 2 of Section 4.4, they are reduced from $O(N^2)$ to $O(N)$. Figure 5.5-1(a) shows how they move in this case when α increases from 0 (as in Figure 4.4-1) to 0.9. Figure 5.5-1(b) shows the effect this has on the accuracy of different Fourier modes. Figure 5.5-2 shows that the smaller DM elements reduce

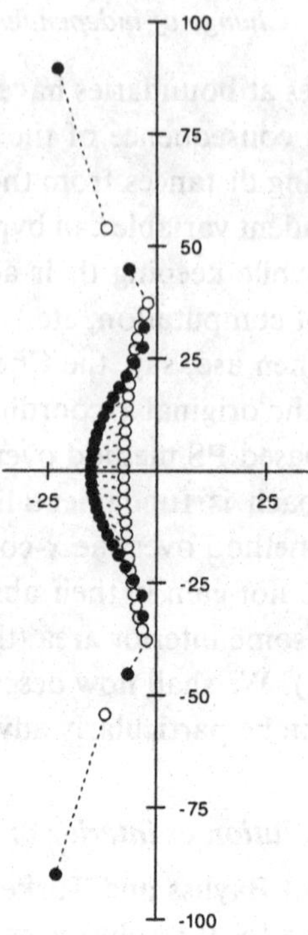
(a)
100
75
50
25
-25
25
-25
-50
-75
-100

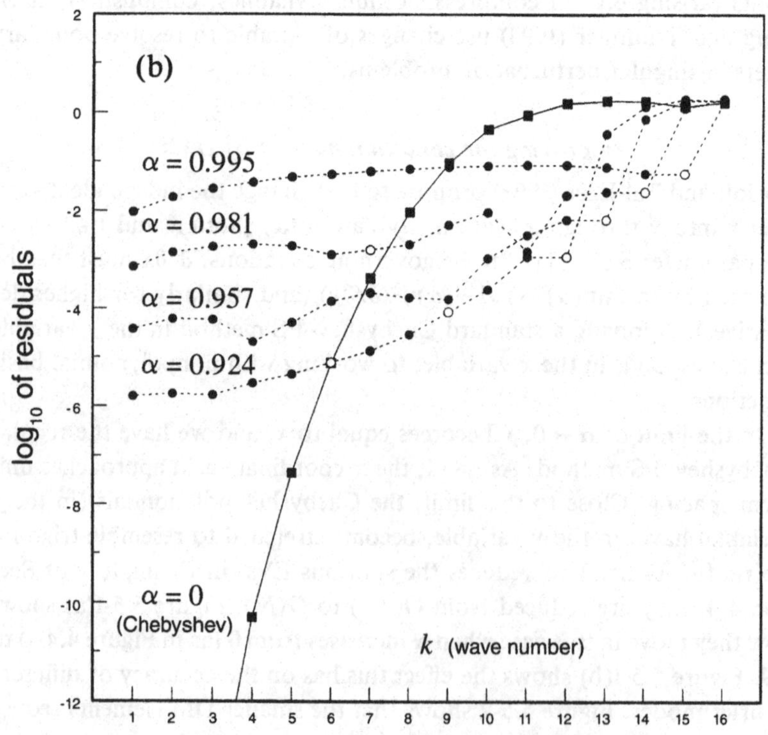
(b)
$\alpha = 0.995$
$\alpha = 0.981$
$\alpha = 0.957$
$\alpha = 0.924$
$\alpha = 0$
(Chebyshev)
$\log_{10}$ of residuals
k (wave number)
2
0
-2
-4
-6
-8
-10
-12
1
2
3
4
5
6
7
8
9
10
11
12
13
14
15
16

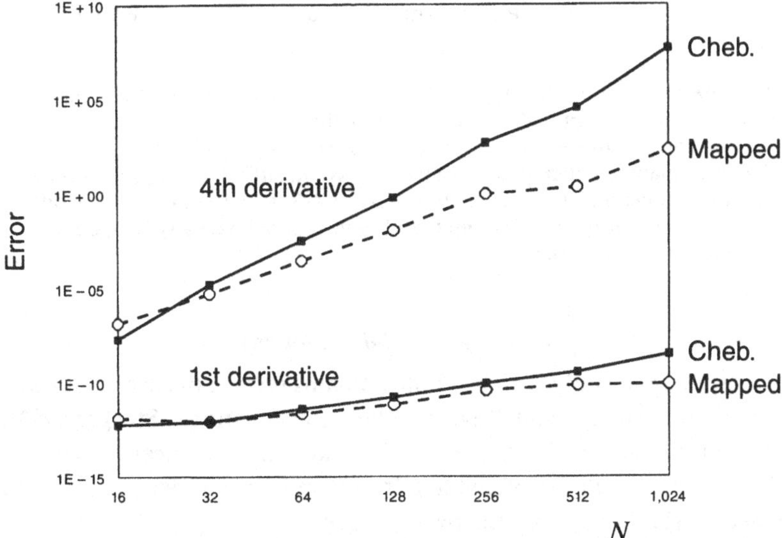

Figure 5.5-2. Size of rounding errors when calculating derivatives of $u(x) = \sin(2x)$, $x \in [-1,1]$. The data are taken from Don and Solomonoff (1995). Their computations were performed (using matrix × vector multiplication) on a Cray C90 with machine precision $\epsilon = 6.5 \cdot 10^{-15}$. The best value for α depends on both ϵ and N, and was selected as follows.

N	16	32	64	128	256	512	1024
α	0.25532	0.63778	0.88252	0.96830	0.99191	0.99797	0.99950

the effect of rounding errors, which are often the dominant error source in cases of smooth functions and large values of N. In contrast, truncation errors decrease exponentially with N.

Compared to the standard Chebyshev–PS method, the mapping procedure offers

- reduced spurious EVs,
- better-conditioned DMs (smaller elements, and DMs that are closer to being normal), and
- a wider range of accurately treated Fourier modes

Figure 5.5-1. Effect on EVs and on accuracy from changes in mapping parameter α ($N = 32$). (**a**) Changes in EVs when α is increased from 0 (Chebyshev case, cf. Figure 4.4-1) to 0.9. (**b**) Residuals when first-derivative approximations are applied to $u_k(x) = \cos(k\pi x)$, $k = 1, 2, \ldots, 16$. [The values of $\alpha = \cos(\pi j/32)$, $j = 1, 2, 3, 4$, correspond to requiring "accuracy" up to node $16 - j$. The approximations become exact in certain cases, indicated by open circles (their vertical positions are artificial and serve only to make the chart easier to read). Numerical data are taken from Kosloff and Tal-Ezer (1993).]

in exchange for a significantly reduced accuracy for the lowest Fourier modes.

The effectiveness of this approach has not yet been carefully compared against, say, medium-to-high-order FD methods.

If we had just moved the gridpoints toward equi-spaced locations without the accompanying change of variable, we would suffer all the problems of the Runge phenomenon (cf. Figures 3.2-2, 3.4-3, and 4.1-1(a)) - disastrous growth of condition number and inability to approximate anything but extremely smooth functions.

Handling of infinite domains

Pseudospectral techniques for infinite domains include the use of Hermite polynomials (Funaro and Kavian 1988), sinc functions (Stenger 1993), (almost) rational functions (Boyd 1987), and domain truncation possibly followed by a change of variable (Grosch and Orszag 1985). Cloot and Weideman (1992) compare the two approaches:

- truncate $[-\infty, \infty]$ to $[-L, L]$ and scale with $y = \pi x/L$ to get $y \in [-\pi, \pi]$, or
- change variable $x = L\tan(y/2)$ to get $y \in [-\pi, \pi]$,

in both cases followed by Fourier-PS approximation. Boyd (1994) analyzes a variation of this in which the Fourier-PS method is preceded by both domain truncation and the change of variable $x = \sinh(Ly)$.

The Hermite-PS method may be briefly summarized as follows. The Hermite polynomials $H_n(x)$ can be obtained through

$$H_0(x) = 1, \quad H_1(x) = 2x, \quad \text{and} \quad H_{n+1}(x) = 2xH_n(x) - 2nH_{n-1}(x),$$

where $n = 1, 2, \ldots$. The Hermite functions $\psi_n(x) = (1/\sqrt{2^n \cdot n!})H_n(x)e^{-x^2/2}$ then become orthogonal over $[-\infty, \infty]$. Derivatives of Hermite expansions can easily be re-expanded as such expansions through use of the relation $\psi_n'(x) = -(\sqrt{(n+1)/2})\psi_{n+1}(x) + (\sqrt{n/2})\psi_{n-1}(x)$. As a consequence of the infinite spatial interval, the distances between nearest nodes (e.g. the zeros of $\psi_{N+1}(x)$) decrease only very slowly with N: like $O(1/\sqrt{N})$ vs. $O(1/N)$ for Fourier-PS and $O(1/N^2)$ for Chebyshev-PS. This tends to lead to favorable stability conditions.

The lack of an efficient discrete Hermite transform suggests numerical implementation through the use of DMs. Concerns about poor convergence rates as N is increased (Gottlieb and Orszag 1977) can be partly overcome through an initial change of variable $y = \alpha x$, where α is a problem-dependent parameter (Tang 1993).

5.6. Domain decomposition and spectral elements

There are many reasons for decomposing a large domain into several smaller ones.

- Simplification of complex geometry – the decomposition can either be fixed in time or be changed dynamically as the geometry and/or the solution evolve.

  Many variations are possible. For example, a rotating irregularly shaped object can be surrounded by a grid with a circular outer edge, allowing an easy connection to a stationary outer grid with the same circular (inner) edge.

- Allowing the use of different resolutions and/or numerical methods in different parts of a domain. (In each subdomain, one can choose a discretization method particularly well suited to local singularities or to other special features that might be present; one can use different time steps; etc.).
- Utilizing the capacity of parallel computers by assigning different domains to different processors.
- Speeding up certain linear algebra tasks by creating operation-saving sparsity structures in global matrices.
- Permitting economical use of methods whose cost increases faster than linearly with domain size.

Global solutions can then be obtained through different strategies, such as

- direct (accelerated) iterations of coupled subdomain solutions,
- combinations of local solutions into a preconditioner for a global iterative process, and/or
- use of local solutions as fine-grid corrections within some type of multigrid environment.

Subdomains can be either overlapping or non-overlapping. In the former case, the *Schwartz alternating method* essentially amounts to obtaining boundary values for each subdomain from the interior solution of another subdomain. When using relatively large overlaps (or when exchanging more information between subdomains – in order to strengthen their coupling – than just boundary values) together with good techniques for convergence acceleration, repeated subdomain solutions can lead to rapid global convergence. A simple and commonly used demonstration problem is to solve Poisson's equation in an L-shaped domain by applying fast Poisson solvers on two rectangular domains that overlap at the bend. Dryja and Widlund (1990) discuss the Schwartz alternating method in a general enough context to provide a link to the alternative approach of non-overlapping domains.

To converge when using direct iterations, multi-domain non-overlapping PS methods need to impose compatibility conditions on both function values and normal derivatives at boundaries (in the case of second-order problems). One way around this is known as the *spectral element* method (introduced in Patera 1984). The idea here is to first recast the governing differential equations into a variational form (reminiscent of finite elements). When using a Galerkin approach with trial functions that are continuous across domain boundaries, the appropriate domain couplings become automatically satisfied. With a fixed number of domains and using spectral representations of increasing accuracies within each of them, global spectral convergence follows. Dubiner (1991) and Mavriplis and van Rosendale (1993) note that spectral basis functions can be found also on triangular subdomains, thus allowing a finite element–like flexibility in the covering of irregular domains.

Since most spectral domain-decomposition techniques use Galerkin formulations, the topic falls somewhat beside the scope of this PS review book. However, the area is important and active. Good sources of papers and references on general domain-decomposition methods are the conference proceedings by Chan et al. (1989, 1990) and the review by Chan and Mathew (1994). For spectral elements, see also Canuto et al. (1988, Chap. 13) and Fischer and Rønquist (1994). Staggered-grid Chebyshev multi-domain methods are discussed in Kopriva and Kolias (1995).

We noted in Sections 3.2 and 3.3 that node clusterings at the ends of an interval could be seen as an undesired measure forced by the need to use one-sided stencils. In Example 3 of Section 5.1 and in Section 6.2 (concerning $r = 0$ in polar coordinates), we observe that severe mesh refinement is wasteful if data somehow can be obtained from outside a boundary. It would therefore seem worthwhile to develop PS domain-decomposition methods where the subdomains overlap (or in some other way share significant amounts of information) and where the nodes are not clustered as densely as for Jacobi polynomials.

6

PS methods in polar and spherical geometries

PDEs in spherical geometries arise in many important areas of application such as meteorology, geophysics, and astrophysics. A fundamental problem for most discretization techniques is that it is impossible to cover a sphere with grids that are both dense and uniform. We begin by describing some numerical approaches designed to address or bypass this problem.

Approximately uniform grids over the sphere. One might start by laying out a coarse, perfectly uniform grid based on one of the five platonic bodies (in particular the icosahedron with 20 equilateral triangular faces), and then carry out subdivisions within each face. Variations on this theme include using grids reminiscent of

- the dimple pattern on golf balls;
- "Buckminster Fullerenes" – or the patterns of carbon atoms in "Buckey balls"; or
- approximations found in biology, such as the pattern of composite eyes in some insects or the silica skeletons of some radiolaria.

Grids of this type can be well suited for low-order FD and FE methods (see e.g. Bunge and Baumgardner 1995), but even their relatively minor irregularities cause considerable algebraic complications in connection with higher-order FD and PS methods.

Spherical (surface) harmonics. These form an infinite set of analytic basis functions with a completely uniform approximation ability over all parts of a sphere. Galerkin techniques are particularly attractive for linear constant-coefficient problems. Equations with variable coefficients and nonlinearities are best handled via (repeated) transformations to and from a grid-based physical representation. Drawbacks include algebraic complexity and lack of very fast transforms.

1-D periodic Fourier–PS approximations applied separately in each direction on a standard spherical polar coordinate grid. This approach is by far the easiest one to implement. However, two potential difficulties have been noted.

CFL problems: The Courant–Friedrichs–Lewy (CFL) stability condition states that an explicit time stepping scheme for a convection-type equation cannot be stable if the geometric shape of the stencil is such that information cannot propagate sideways as fast as it does in the governing equation. The CFL condition implies an upper limit on k/h. Tiny values of h near the poles therefore severely restrict the time steps k that can be used. For a discussion on the CFL condition, see e.g. Strang (1986, p. 579).

Parity problems: Longitude and latitude circles on a sphere intersect at two places. A Fourier mode in one of the directions may conflict with certain modes in the other direction. In 2-D Fourier expansions, only certain mode combinations are permissible. This issue is discussed at length in Boyd (1989).

When using PS approximations, derivatives in different spatial directions can be approximated entirely independently of each other. This independence is taken for granted (and is always exploited) when designing FD approximations. Consequently, in each direction, we consider here the 1-D FD schemes of maximal accuracy – that is, Fourier- or Chebyshev-type PS approximations. We will later see how a simple smoothing strategy can bypass the CFL problem (while ensuring a uniform resolution over the sphere).

Truncated trigonometric expansions are unique in featuring exactly the same resolution ability at all locations within their period (the trigonometric basis functions cos and sin need not be postulated, but can be derived from this requirement). Section 6.1 outlines how spherical harmonics can be similarly derived from a corresponding requirement of achieving exactly identical resolution over all parts of the surface of a sphere. Many books in applied mathematics and quantum mechanics describe spherical harmonics (see e.g. Hobson 1931). Numerically oriented discussions can be found for example in Orszag (1974) and Boyd (1989, Chap. 15).

In Section 6.2, we discuss the direct use of 1-D PS approximations in the case of polar coordinates in the plane. In Section 6.3, we apply this same idea to equations given on the surface of a sphere. Example 1 (purely convective flow over the surface of a sphere) does not indicate any problems associated with grid clustering near the poles. This PS method for a sphere is found to clearly outperform low-order FD methods in the chosen model problem (designed to test only the issue of polar grid singularities, not the range of additional complications that arise in applications such as meteorology).

For most flow problems in spherical geometries (e.g. the Navier–Stokes equations), one or more elliptic equation(s) must be solved at each time step. Example 2 in Section 6.3 shows that Poisson's equation on a sphere can already be solved very accurately with low-order FD methods. This suggests pairing PS methods for the convective equations with FD methods for the elliptic ones.

Sections 6.2 and 6.3 expand on the discussion in Fornberg (1995). The polar and spherical procedures can be combined to form an entirely FFT-based Fourier- or Chebyshev–PS method for the interior of a sphere.

6.1. Spherical harmonics

In Section 2.1, we quoted trigonometric polynomials as the obvious way to expand periodic functions. A truncated trigonometric expansion never needs any additional frequency components if the independent variable is shifted by any constant amount; only the values of existing expansion coefficients will change. Viewed as defined on the periphery of a unit circle, truncated Fourier expansions form a closed set under rotations.

Trigonometric expansions of increasing order are conceptually related to successively refined uniform grids along the periphery of the unit circle. Because uniform grids on the surface of the sphere can have no more than 20 points, it is not intuitively obvious that any infinite sequence of uniformly accurate expansion functions can exist in this case. However, one can always try to search for some, and a good starting point is to consider Laplace's equation

$$\frac{\partial^2 u}{\partial x^2}+\frac{\partial^2 u}{\partial y^2}+\frac{\partial^2 u}{\partial z^2}=0. \tag{6.1-1}$$

This equation can easily be shown to be invariant under any rotation (and translation) of the (x, y, z)-coordinate frame. We may then hope that the eigenvalue problem based on the Laplacian operator will give rise to eigenfunctions which, in some sense, also are rotation invariant. To explore this possibility, we proceed by introducing spherical polar coordinates through the change of variables

$$\begin{aligned} x &= r\cos\theta\cos\varphi, \\ y &= r\cos\theta\sin\varphi, \\ z &= r\sin\theta. \end{aligned} \tag{6.1-2}$$

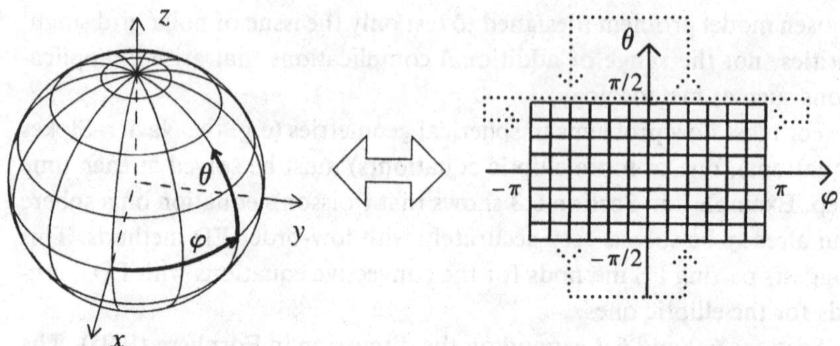

Figure 6.1-1. Grid arrangement and periodicities for spherical coordinates.

See Figure 6.1-1.

In order to work with symmetric domains as far as possible ($\theta \in [-\pi/2, \pi/2]$, $\varphi \in [-\pi, \pi]$; cf. $x, y, z \in [-1, 1]$ for the unit sphere), and also to be more consistent with the latitude–longitude grid on the earth, we measure θ from the equator (latitude) rather than down from the north pole (co-latitude). To swap between the two definitions, one simply exchanges $\sin\theta$ and $\cos\theta$, $\tan\theta$ and $\cot\theta$, and $\partial/\partial\theta$ and $-\partial/\partial\theta$.

Equation (6.1-1) then becomes

$$r^2\frac{\partial^2 u}{\partial r^2}+2r\frac{\partial u}{\partial r}+\frac{\partial^2 u}{\partial \theta^2}-(\tan\theta)\frac{\partial u}{\partial \theta}+\frac{1}{\cos^2\theta}\frac{\partial^2 u}{\partial \varphi^2}=0. \qquad (6.1\text{-}3)$$

We limit our interest to functions without any r dependence, and let the domain be the surface of the unit sphere (i.e. $r=1$). Consider next the eigenvalue problem

$$\nabla^2 u = -n(n+1)u$$

with the surface Laplacian operator (as (6.1-3), but with the r derivatives removed)

$$\nabla^2=\frac{\partial^2}{\partial \theta^2}-(\tan\theta)\frac{\partial}{\partial \theta}+\frac{1}{\cos^2\theta}\frac{\partial^2}{\partial \varphi^2}. \qquad (6.1\text{-}4)$$

Separation of variables can now be used. Setting $u(\theta,\varphi)=P(z)\cdot\Phi(\varphi)$ (where $z=\sin\theta$; cf. equation (6.1-2)) leads to the pair of equations

$$\Phi''+m^2\Phi=0, \qquad (6.1\text{-}5)$$

$$(1-z^2)\frac{\partial^2 P}{\partial z^2}-2z\frac{\partial P}{\partial z}+\left[n(n+1)-\frac{m^2}{1-z^2}\right]P=0. \qquad (6.1\text{-}6)$$

The only 2π-periodic solutions to (6.1-5) are

Table 6.1-1. *The polynomials* $(d^m/dz^m)P_n(z)$ *for* $m = 0, 1, \ldots, 4$ *and* $n = 0, \ldots, m$

n	m: 0	1	2	3	4	$\ldots$
0	1					
1	z	1				
2	$\frac{3}{2}z^2 - \frac{1}{2}$	$3z$	3			
3	$\frac{5}{2}z^3 - \frac{3}{2}z$	$\frac{15}{2}z^2 - \frac{3}{2}$	$15z$	15		
4	$\frac{35}{8}z^4 - \frac{15}{4}z^2 + \frac{3}{8}$	$\frac{35}{2}z^3 - \frac{15}{2}z$	$\frac{105}{2}z^2 - \frac{15}{2}$	$105z$	105	
$\vdots$	$\vdots$	$\vdots$	$\vdots$	$\vdots$	$\vdots$	$\ddots$

Note: When multiplied with $(1-z^2)^{m/2}$, these polynomials form the associated Legendre functions $P_n^m(z)$.

$$\Phi^m(\varphi) = \begin{cases} \cos m\varphi, \\ \sin m\varphi, \end{cases} \qquad m = 0, 1, 2, \ldots.$$

For $m = 0$, (6.1-6) defines the Legendre polynomials (cf. Table A-1). For $m \geq 0$, we obtain the *associated Legendre functions,* which are expressible in terms of the Legendre polynomials as follows:

$$P_n^m(z) = (1-z^2)^{m/2} \frac{d^m}{dz^m} P_n(z), \quad n = m, m+1, m+2, \ldots.$$

Equation (6.1-6) possesses an additional set of solutions, often denoted by $Q_n^m(z)$. They have logarithmic singularities at $z = \pm 1$, and are not of immediate interest in the present context.

Table 6.1-1 shows $(d^m/dz^m)P_n(z)$ for selected values of m and n (with the first column being the Legendre polynomials, as listed in Table A-1). Several closed-form expressions are available for the polynomials in this table, including

$$\frac{d^m}{dz^m} P_n(z) = \frac{1}{2^n \cdot n!} \frac{d^{m+n}}{dz^{m+n}} (z^2 - 1)^n \qquad \text{(Rodrigues's formula)}$$

$$= \frac{1}{2^n} \sum_{k=0}^{[(1/2)(n-m)]} \frac{(-1)^k (2n-2k)!}{k!\,(n-k)!\,(n-m-2k)!} z^{n-m-2k}$$

$$= \frac{(2m)!}{2^m \cdot m!} C_{n-m}^{m+1/2}(z).$$

In the last case, the Gegenbauer (or "ultraspherical") polynomials

$$C_n^{(\alpha)}(z)=\frac{\Gamma(\alpha+\frac{1}{2})\Gamma(2\alpha+n)}{\Gamma(2\alpha)\Gamma(\alpha+n+\frac{1}{2})}P_n^{(\alpha-1/2,\,\alpha-1/2)}(z)$$

are special cases of the Jacobi polynomials introduced in Section 2.1 and Appendix A (with a different normalization). More important (from a numerical point of view) than these closed-form expressions are the many recursion relations that allow rapid evaluation of $P_n^m(z)$, $(d/dz)P_n^m(z)$, and so forth for different values of m and n (see Belousov 1962). Examples include

$$(2n+1)zP_n^m(z)=(n-m+1)P_{n+1}^m(z)+(n+m)P_{n-1}^m(z)$$

and

$$(1-z^2)\frac{d}{dz}P_n^m(z)=(n+1)zP_n^m(z)-(n-m+1)P_{n+1}^m(z).$$

The key remaining question is how to truncate the doubly infinite sequences $m=0,1,2,\dots$ and $n=m,m+1,m+2,\dots$ so that, for each level of truncation, the result forms a closed set with respect to arbitrary rotations. The answer turns out to be very simple: For any value of n, we need to include $m=0,1,2,\dots,n$ (i.e., always truncate after a full row in Table 6.1-1 and in Figure 6.1-2; for a proof, see e.g. Merzbacher 1970).

The associated Legendre functions are often normalized to $S_n^m(z)=(\sqrt{(2n+1)(n-m)!/4\pi(n+m)!}\,)P_n^m(z)$ in order to satisfy $\int_{-1}^{1}[S_n^m(z)]^2\,dz=1/2\pi$, $n\ge m\ge 0$. For any truncation level N, the form of the (surface) spherical harmonics expansion thus becomes

$$u(\varphi,z)=\sum_{n=0}^{N}\Bigg[a_{0,n}S_n^0(z)+\sum_{m=1}^{n}(a_{m,n}S_n^m(z)\cos m\varphi \\ +b_{m,n}S_n^m(z)\sin m\varphi)\Bigg]. \quad (6.1\text{-}7)$$

Since the basis functions are orthonormal, this expansion shares many characteristics with standard trigonometric expansions. For example, least-squares approximations of increasing orders of accuracy are obtained by adding terms without any alterations of previous coefficients. Also, an immediate generalization of Parseval's relation for 1-D and 2-D Fourier expansions is available.

Figure 6.1-2 illustrates how the oscillation pattern of spherical harmonic basis functions $S_n^m(z)\cos m\varphi$ evolves as n and m increase. Figure 6.1-3 also shows the numerical values of the basis functions when $n=12$ and $m=0$, 4, 8, and 12. For fixed n and increasing m, the main peaks shift from the polar to the equatorial regions – all precisely balanced so

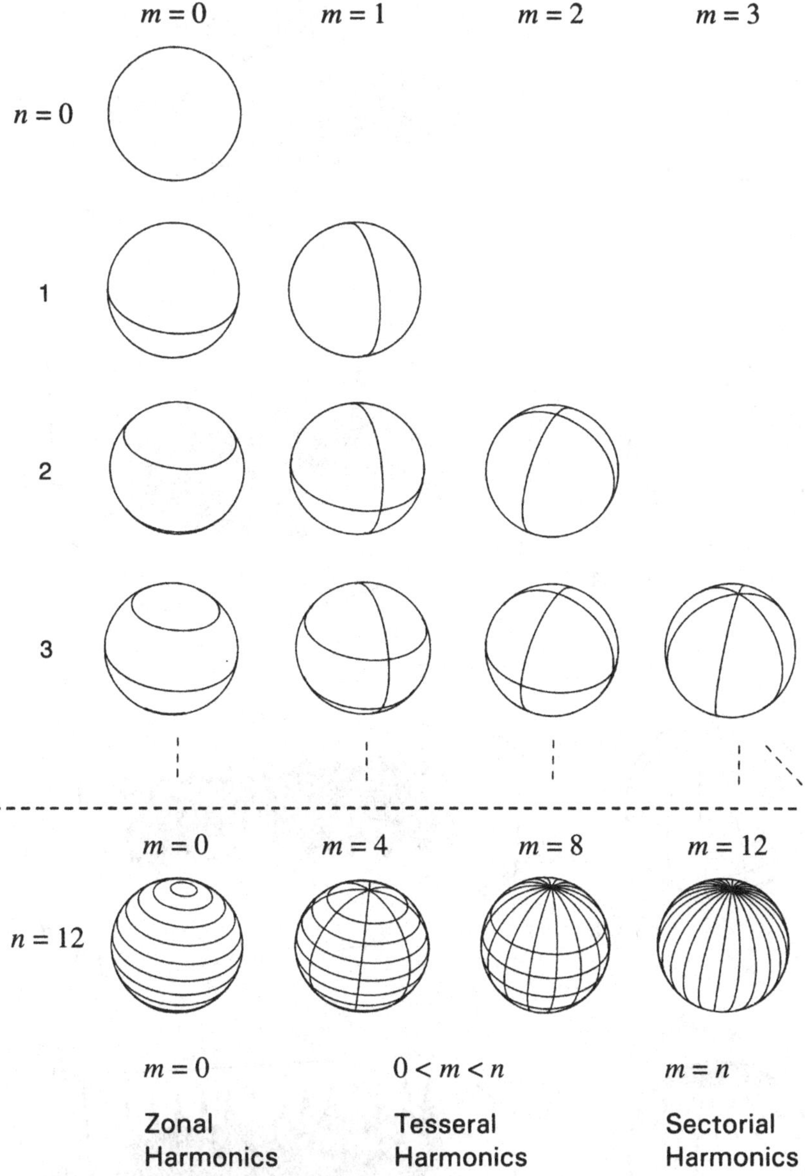

Figure 6.1-2. Zero contours (curves of sign changes) for some low-order spherical harmonics.

that the net result, at any truncation level, is an exact uniformity of approximation over the complete sphere.

In spite of their mathematical elegance, the attractiveness of spherical harmonics in numerical contexts is limited because of their high algebraic

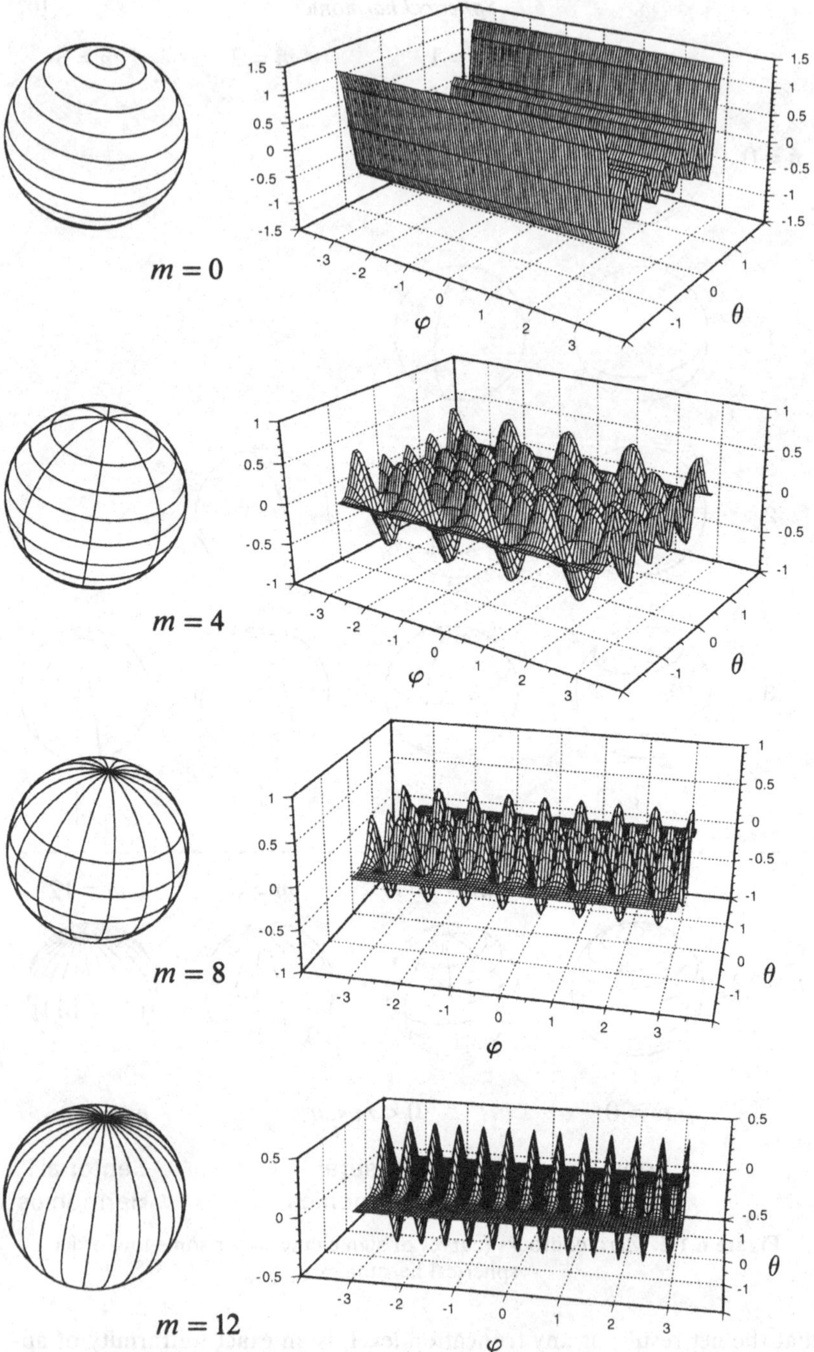

Figure 6.1-3. Spherical harmonic basis functions $S_n^m(z) \cos m\varphi$ for $n = 12$, $m = 0, 4, 8, 12$. Displays show curves of sign changes on the unit sphere and numerical values in the (φ, θ) plane.

complexity, and because fast conversions between coefficients for an expansion in $P_n^m(z)$, $n = 0, 1, 2, \ldots, N$, and function values at a set of gridpoints z_j, $j = 0, 1, \ldots, N$, have been found only for $m = 0$.

In the $m = 0$ case, Legendre expansions can be converted to and/or from Chebyshev expansions by a multi-pole method (Alpert and Rokhlin 1991) or by a wavelet method (Beylkin, Coifman, and Rokhlin 1991; improved in Beylkin and Brewster 1995). Chebyshev expansions can in turn be converted to and from gridpoint values by the fast cosine transform. For both approaches, we note first that if $\sum_{k=0}^{N} a_k T_k(x) = \sum_{k=0}^{N} b_k P_k(x)$, then

$$\begin{bmatrix} 1 & 0 & -\frac{1}{3} & 0 & -\frac{1}{15} & 0 & -\frac{1}{35} & \cdots \\ & 1 & 0 & -\frac{3}{5} & 0 & -\frac{1}{7} & 0 & \cdots \\ & & \frac{4}{3} & 0 & -\frac{16}{21} & 0 & -\frac{4}{21} & \cdots \\ & & & \frac{8}{5} & 0 & -\frac{8}{9} & 0 & \cdots \\ & & & & \frac{64}{35} & 0 & -\frac{384}{385} & \cdots \\ & & & & & \frac{128}{63} & 0 & \cdots \\ & & & & & & \frac{512}{231} & \cdots \\ & & & & & & & \ddots \end{bmatrix} \begin{bmatrix} a_0 \\ a_1 \\ a_2 \\ \vdots \\ \vdots \\ \vdots \\ \vdots \\ a_N \end{bmatrix} = \begin{bmatrix} b_0 \\ b_1 \\ b_2 \\ \vdots \\ \vdots \\ \vdots \\ \vdots \\ b_N \end{bmatrix},$$

and similarly for the inverse transform (in that case, all nonzero elements are positive). Simple recursions as well as closed-form expressions are available for the elements. Assuming N to be odd, the preceding matrix separates into two upper triangular full matrices,

$$\begin{bmatrix} \bullet & \bullet & \bullet & \cdots & \bullet \\ & \bullet & \bullet & \cdots & \bullet \\ & & \bullet & \cdots & \bullet \\ & & & \ddots & \vdots \\ & & & & \bullet \end{bmatrix} \begin{bmatrix} a_0 \\ a_2 \\ a_4 \\ \vdots \\ a_{N-1} \end{bmatrix} = \begin{bmatrix} b_0 \\ b_2 \\ b_4 \\ \vdots \\ b_{N-1} \end{bmatrix} \quad \text{and} \quad \begin{bmatrix} \bullet & \bullet & \bullet & \cdots & \bullet \\ & \bullet & \bullet & \cdots & \bullet \\ & & \bullet & \cdots & \bullet \\ & & & \ddots & \vdots \\ & & & & \bullet \end{bmatrix} \begin{bmatrix} a_1 \\ a_3 \\ a_5 \\ \vdots \\ a_N \end{bmatrix} = \begin{bmatrix} b_1 \\ b_3 \\ b_5 \\ \vdots \\ b_N \end{bmatrix},$$

and similarly for the inverse transform. In the multi-pole approach, the idea is to split each of the matrices into a block form, with the smallest blocks along the diagonal and increasingly larger blocks toward the top right. A smooth variation of the coefficients according to their position allows each block to be accurately represented by a low-rank matrix.

In the wavelet approach, the linear unitary transform corresponding to a change to a wavelet coefficient basis makes most of the matrix elements extremely small; again, the smoothness of the original entries away from the diagonal allows for a more compact representation. If elements below (say) $\epsilon = 10^{-7}$ or $\epsilon = 10^{-15}$ are discarded, then only a few narrow bands of elements remain.

Both approaches can be executed in $O(N(-\log \epsilon))$ operations, as compared to $O(N \log N)$ for a fast cosine transform - the sizes of the constants in the O symbols are comparable. In the case of associated Legendre functions, both of these approaches fail because the matrix entries no longer vary smoothly with their positions (even far from the diagonal).

Dilts (1985) presents an algorithm that converts more effectively between equi-spaced (φ, θ) data and spherical harmonic expansion coefficients than do Gaussian quadrature-type methods. On modern computers, direct matrix × vector multiplication is often still more effective.

Of the three requirements for expansion functions listed at the beginning of Section 2.1, (1) and (2) are better met by spherical harmonic functions than is requirement (3).

6.2. PS approximations in polar coordinates

A polar coordinate system on the unit circle can be obtained through the transformation $x = r\cos\theta$, $y = r\sin\theta$.

At $r = 0$, all the θ positions collapse into one physical gridpoint – therefore requiring only one governing equation. At this location, one can use a Cartesian (x, y)-based FD stencil (free of the singularities that might have been introduced by the change to polar coordinates).

When using Fourier–PS approximations in θ and Chebyshev–PS approximations in r, it is important to consider not the conventional polar domain $0 \le r \le 1$, $-\pi \le \theta \le \pi$, but instead $-1 \le r \le 1$, $-\pi/2 \le \theta \le \pi/2$. This makes it unnecessary to refine the grid in the r direction near $r = 0$, with several consequential benefits:

- reduced number of gridpoints;
- higher-order accuracy of PS stencils in the r directions (since they extend over twice as many gridpoints – cf. Example 3 in Section 5.1); and
- less severe 2-D point clustering near the origin.

Other points to note include:

- smoothing in the θ direction for small r values enables a favorable CFL stability condition (without damaging overall accuracy); and
- availability of the Fourier–PS method (as before) in the θ direction.

Example 3 in Section 5.1 illustrated (in an antisymmetric case) this same idea of letting FD stencils extend right over the pole.

In cases of axisymmetry, the present method may superficially resemble a proposal in Canuto et al. (1988, p. 90) to expand in r using only even-order Chebyshev polynomials. However, that approach fails as a PS procedure if axisymmetry is lost (p. 91); in contrast, the present approach carries immediately over. This makes possible the pole treatment in the case of spherical coordinates discussed in Section 6.3.

6.3. The PS method in spherical coordinates

We consider again the surface (φ, θ) grid shown in Figure 6.1-1. The dotted arrows in the right part indicate how periodicity can be implemented in both φ and θ. The PS approximations for the derivatives in the two directions can be approximated entirely separately, as for the case of independent 1-D problems. An r direction can be added in the obvious way. If $r = 0$ is included in the domain of interest, we can again consider $-1 \le r \le 1$ and halve the angular extent. The simple model problem discussed next is designed solely to show that the grid singularities at the poles need not cause any difficulties for convective flow problems.

Example 1. Compare FD and PS simulations of purely convective flow in different directions over the surface of a unit sphere.

Figure 6.3-1 shows two simple flow situations that suffice to illustrate the effect (if any) of polar coordinate singularities: solid body–type convection around a sphere in the direction of the equator (Case 1) and in a direction across the polar regions (Case 2). Other flow patterns over the sphere will correspond to different choices of the variable coefficients $a(\varphi, \theta)$ and $b(\varphi, \theta)$. As in Case 2, $a(\varphi, \pm\pi/2)$ will typically be infinite (a consequence of the grid singularity at the poles).

The FD2, FD4, and Fourier–PS methods were implemented in an entirely straightforward manner (taking note of the periodicities indicated in Figure 6.1-1). With regard to the poles, $u(\varphi, \pm\pi/2)$ was left unchanged throughout the time stepping in Case 1, and the Cartesian equation

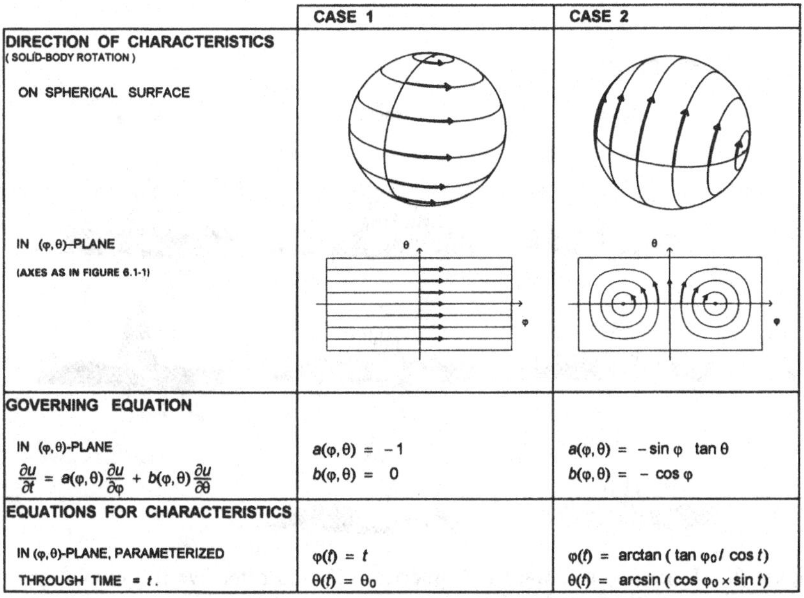

	CASE 1	CASE 2
DIRECTION OF CHARACTERISTICS (SOLID-BODY ROTATION) ON SPHERICAL SURFACE IN (φ, θ)-PLANE (AXES AS IN FIGURE 6.1-1)	θ φ	θ φ
GOVERNING EQUATION IN (φ, θ)-PLANE $\frac{\partial u}{\partial t} = a(\varphi, \theta)\frac{\partial u}{\partial \varphi} + b(\varphi, \theta)\frac{\partial u}{\partial \theta}$	$a(\varphi, \theta) = -1$ $b(\varphi, \theta) = 0$	$a(\varphi, \theta) = -\sin\varphi \tan\theta$ $b(\varphi, \theta) = -\cos\varphi$
EQUATIONS FOR CHARACTERISTICS IN (φ, θ)-PLANE, PARAMETERIZED THROUGH TIME = t.	$\varphi(t) = t$ $\theta(t) = \theta_0$	$\varphi(t) = \arctan(\tan\varphi_0 / \cos t)$ $\theta(t) = \arcsin(\cos\varphi_0 \times \sin t)$

Figure 6.3-1. Example 1: Two test cases for convective flow over a sphere.

$$\left.\frac{\partial u}{\partial t}\right|_{\text{all }\varphi} = -\left.\frac{\partial u}{\partial \theta}\right|_{\varphi=0}$$

was used at $\theta = \pm\pi/2$ in Case 2.

No numerical smoothing was applied in the θ direction in any of the test runs. (However, a small amount of damping of the very highest-frequency components is probably advisable to enhance the robustness of production codes.) In the φ direction, an FFT-based smoothing was performed after each time step. The proportion $1-\cos\theta$ of the modes was cut out. This removal of the highest modes (i.e., none removed at the equator, half at $\theta = \pm\pi/3$) leads to precisely the same resolution power of the calculation at all parts of the sphere, and bypasses the otherwise restrictive polar CFL stability condition. In the FD cases, much simpler FD-type smoothing operators could quite certainly have been used with equally good results.

Figure 6.3-2 shows the numerical solutions at time $t = 10\pi$ (i.e. after five full revolutions) in Case 1, and Figure 6.3-3 shows the same for Case 2. The initial condition was a circular cone in the (φ, θ) plane. Time stepping was carried out with a standard, fixed-step, fourth-order Runge-Kutta scheme. The time steps were sufficiently small that the visible errors are all due to the spatial discretization. With the PS method in space, the size of the time steps was limited by accuracy (rather than stability) considerations.

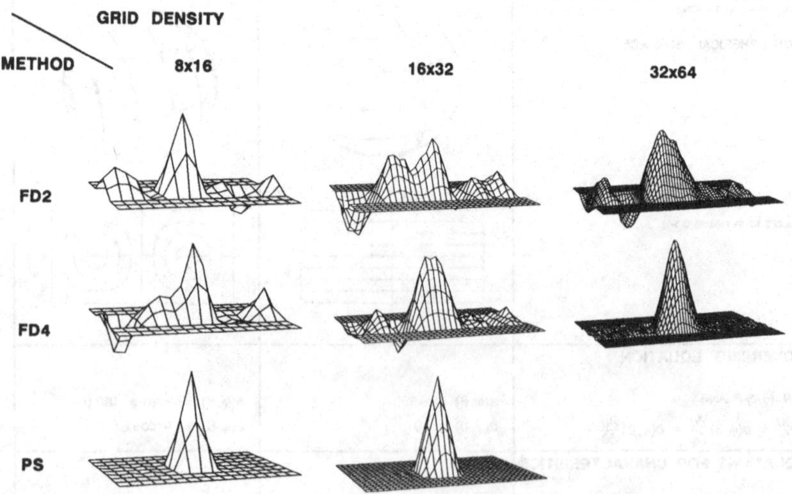

Figure 6.3-2. Example 1, Case 1: Numerical solution after five full revolutions using different methods and grid densities. The (φ, θ)-plane grid layout is as shown in Figure 6.1-1.

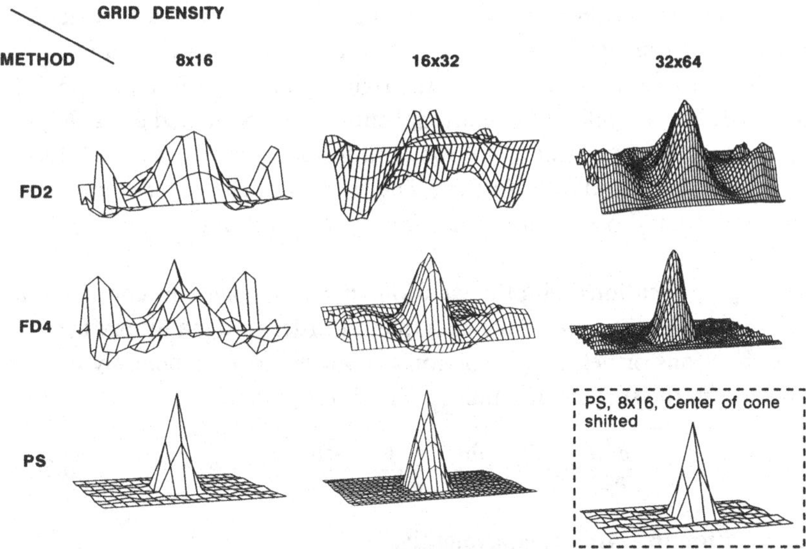

Figure 6.3-3. Example 1, Case 2: Numerical solution after five full revolutions using different methods and grid densities. The insert at the bottom right shows the same case as bottom left, but with the cone shifted one grid point to the right (causing it to pass near the poles rather than across them).

Case 1 is trivial in the sense that the governing equation separates into independent, identical, constant-coefficient 1-D equations for the different θ positions. The PS method happens to become exact in this case (apart from the smoothing, which has negligible influence). The errors we see in Figure 6.3-2 are as expected from similar, earlier tests (Fornberg 1988a; see also Gottlieb and Orszag 1977, pp. 135–8). This case represents the highest level of performance these methods are capable of under the most favorable circumstances.

Case 2 and Figure 6.3-3 are notable in that none of the methods has suffered significantly from the fact that the convection has taken place across the polar regions. In particular, the PS method has retained its superior accuracy in spite of strongly variable coefficients of the governing equation. To illustrate that convecting the cone exactly straight across the poles does not constitute an especially favorable test case (because e.g. of computational symmetry in the φ direction), the insert at the bottom right in Figure 6.3-3 shows the result when the initial cone was displaced one gridpoint to the right. After 10 near-pole passages, the result remains equally good.

The computational cost for the FD methods scale roughly as N^3, where N is the number of gridpoints in each spatial direction (taking into account a time-step stability restriction $k \cdot N <$ const.). The cost of a PS calculation is about four times that of the FD case above it, but one half of the FD case above and one column to the right – as laid out in Figures 6.3-2 and 6.3-3. For a similar computational effort, the PS method gives clearly superior accuracy. In full 3-D flows (e.g. inside a sphere), these scalings favor the PS method even further. The savings in computer storage are larger still than those realized for number of operations.

In many applications, elliptic equations must be solved in conjunction with the convective ones – for example, in order to dynamically update flow directions or velocities. Poisson's equation arises commonly in this context. On the surface of a unit sphere, it takes the form

$$\frac{\partial^2 u}{\partial \theta^2} - (\tan\theta)\frac{\partial u}{\partial \theta} + \frac{1}{\cos^2\theta}\frac{\partial^2 u}{\partial \varphi^2} = f(\varphi, \theta) \tag{6.3-1}$$

(cf. equation (6.1-4)), or, equivalently,

$$\frac{1}{\cos\theta}\frac{\partial}{\partial\theta}\left((\cos\theta)\frac{\partial u}{\partial\theta}\right) + \frac{1}{\cos^2\theta}\frac{\partial^2 u}{\partial\varphi^2} = f(\varphi, \theta). \tag{6.3-2}$$

The solution $u(\varphi, \theta)$ is unique (except with respect to an arbitrary additive constant), on the condition that

$$\int_{-\pi}^{\pi}\int_{-\pi/2}^{\pi/2} (\cos\theta) f(\varphi, \theta)\, d\varphi\, d\theta = 0. \tag{6.3-3}$$

This condition (6.3-3) follows from multiplying (6.3-2) by $\cos\theta$ and integrating over φ and θ; the left-hand side then vanishes.

The left-hand side of (6.3-2) can be approximated to second-order accuracy with the FD stencil

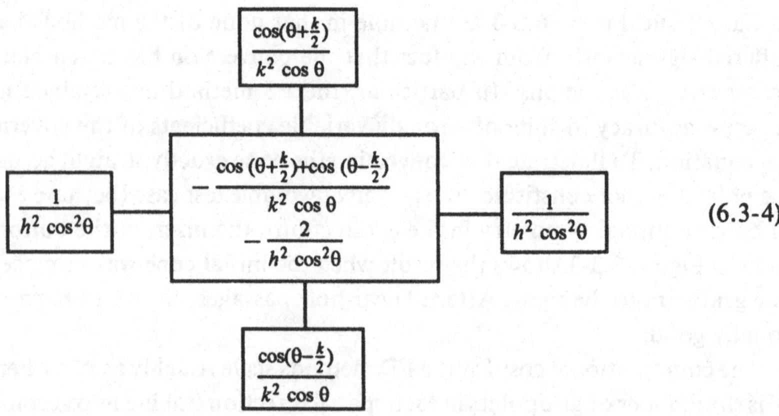

(6.3-4)

at all locations apart from centered at the poles. At the poles, we use instead

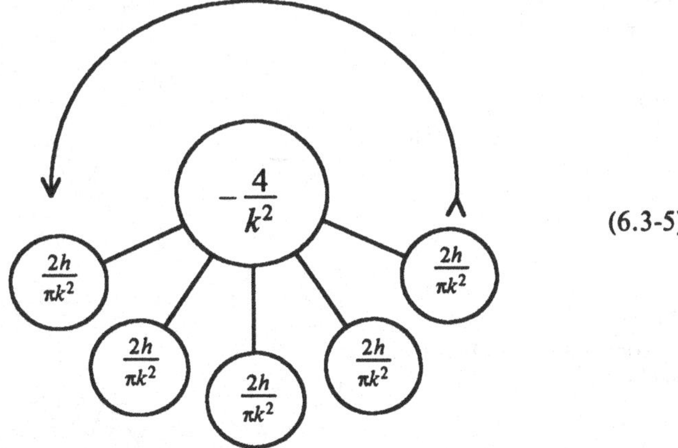

(6.3-5)

In these approximations, h and k denote the grid spacings in the φ and θ directions, respectively.

Instead of (6.3-5), we could have used a Cartesian five-point stencil at the poles (as in Example 1). However, the present average over rotated five-point stencils considerably simplifies the following equation (6.3-6).

The FD2 approximation just described features diagonal (semi-)dominance: all off-center weights are positive. The condition (6.3-3) corresponds to

$$\frac{\pi}{2h}\sin\left(\frac{k}{2}\right)\left[f\left(\cdot,\frac{\pi}{2}\right)+f\left(\cdot,-\frac{\pi}{2}\right)\right]+\sum_{\text{all gridpoints}}(\cos\theta)f(\varphi,\theta)=0. \tag{6.3-6}$$

Numerically obtained data for $f(\varphi,\theta)$ is unlikely to satisfy (6.3-6) exactly. If the expression takes the value ϵ (rather than 0), subtracting the $O(\epsilon hk)$ quantity $(\epsilon h/\pi)/[2\cot(k/2)+\sin(k/2)]$ from f at each gridpoint restores (6.3-6).

An FFT-based direct solution method is available for the resulting discrete system (Swarztrauber 1974). Iterative procedures can also be used – such methods as Jacobi, Gauss–Seidel, SOR, and multi-grid all converge very rapidly in this case (and generalize easily to other elliptic equations).

Example 2. Solve Poisson's equation on a sphere by a second-order FD approximation.

Figure 6.3-4(a) shows, on a very coarse grid, a function $f(\varphi,\theta)$ that satisfies both (6.3-3) and (6.3-6) exactly. Part (b) of the figure shows the approximate solution $u(\varphi,\theta)$ obtained by iteration based on (6.3-4) and (6.3-5). Part (c) shows the difference between the numerical and the analytic solution of this problem.

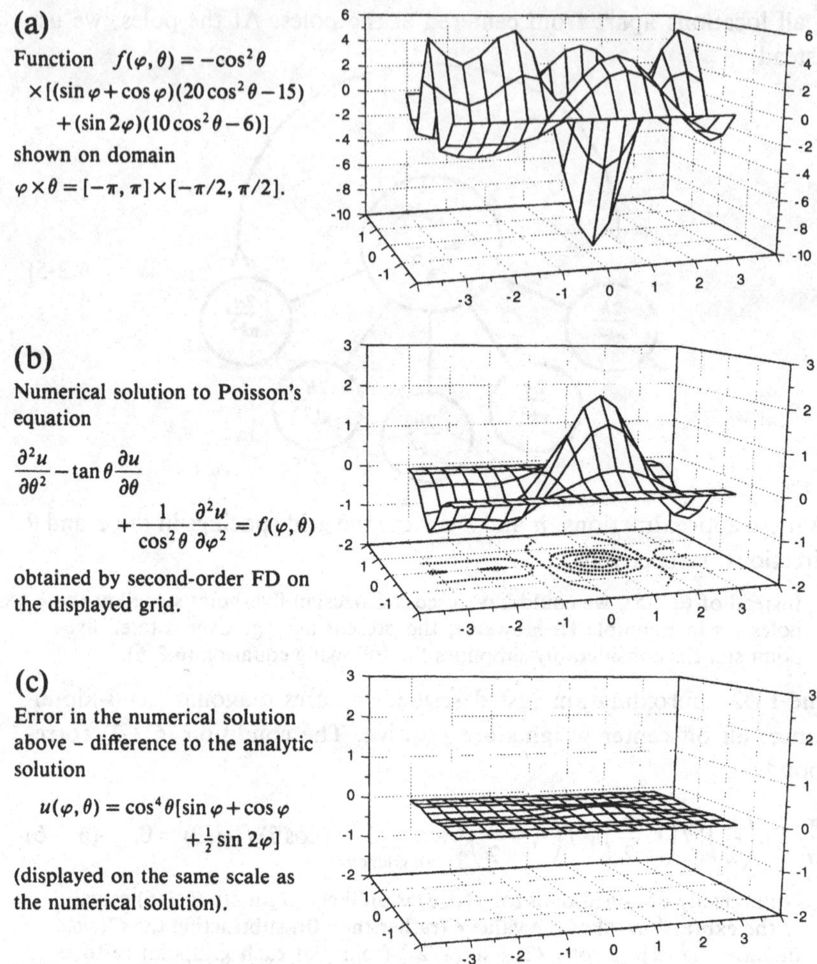

(a)
Function $f(\varphi,\theta) = -\cos^2\theta$
$\times[(\sin\varphi+\cos\varphi)(20\cos^2\theta-15)$
$+(\sin 2\varphi)(10\cos^2\theta-6)]$
shown on domain
$\varphi\times\theta = [-\pi,\pi]\times[-\pi/2,\pi/2]$.
(b)
Numerical solution to Poisson's equation
$\frac{\partial^2 u}{\partial\theta^2} - \tan\theta\frac{\partial u}{\partial\theta} + \frac{1}{\cos^2\theta}\frac{\partial^2 u}{\partial\varphi^2} = f(\varphi,\theta)$
obtained by second-order FD on the displayed grid.
(c)
Error in the numerical solution above – difference to the analytic solution
$u(\varphi,\theta) = \cos^4\theta[\sin\varphi+\cos\varphi + \frac{1}{2}\sin 2\varphi]$
(displayed on the same scale as the numerical solution).

Figure 6.3-4. Example 2: Solution of Poisson's equation on a sphere using second-order finite differences.

Example 2 was selected to demonstrate that, even with a very coarsely represented function, we must lower the FD order all the way down to 2 to enable visual location of any errors in a (periodic) elliptic solution. In a typical flow problem, the contour curves in Figure 6.3-4(b) correspond to streamlines. The small errors in the solution would cause the fluid particles to follow slightly perturbed trajectories; that is, these errors do not have consequences even remotely comparable to the total breakdown of the FD methods on coarse grids seen for the convective equations in Figures 6.3-2 and 6.3-3.

If – in some time-dependent flow problem – it is found that the elliptic solvers contribute more to the global error than does the time stepping of the hyperbolic equation(s), the order of the former should be increased to 4 (or still higher). Many approaches for this are available:

- iterative improvement – second-order FD preconditioning of higher-order FD approximations;
- compact high-order schemes – described as *Mehrstellenverfahren* by Collatz (1960) (recent references can be found in Gupta 1991); and
- spectral solvers – for example, Yee (1981) describes a direct FFT-based Galerkin spectral solver for Poisson's equation on a sphere.

7

Comparisons of computational cost for FD and PS methods

High-order FD and PS methods are particularly advantageous in cases of

- high smoothness of solution (but note again the discussion in Section 4.2),
- stringent error requirement,
- long time integrations, and
- more than one space dimension.

Because the PS methods for periodic and nonperiodic problems are quite different, the two cases are discussed separately in what follows. In both cases, we find that the PS methods compare very favorably against FD methods in simple model situations. However, in cases with complex geometries or severe irregularities in the solutions, lower-order FD (or FE) methods may be both more economical and more robust.

Especially for nonperiodic problems, it can be difficult to estimate a priori the computational expense required to solve a problem to a desired accuracy. Many implementation variations are possible, and the optimal selection of formal orders of accuracy, level of grid non-uniformity, and so forth may well turn out to depend not only on the problem type, but also on the solution regimes that are studied. Therefore, it makes sense to keep open as many of these implementation options as possible while developing application codes. One technique is to write an FD code of variable order of accuracy on a grid with variable density (using the algorithm in Section 3.1 and Appendix C). By simply changing parameter values, one can then explore (and exploit) the full range of methods from low-order FD on a uniform grid to Chebyshev (Legendre, etc.) and other PS methods. Obviously, it is also desirable to structure codes so that time stepping methods (if present) are easily interchangeable.

Comparisons of cost versus performance for FD and PS methods are given in Section 7.1 for a periodic and in Section 7.2 for a nonperiodic demonstration problem; see also Section 6.3. A more realistic comparison (FD4 and FD6 vs. spherical harmonics-based PS) is quoted in Section 8.3. Examples of how the accuracy (not the cost) of approximations increases with their order (FD2 → FD4 → ⋯ → PS) can be found in several places (see e.g. Sections 4.7, 5.1, 8.2, and 8.4). Although many more examples can easily be constructed, the design of a comprehensive, convenient, and sharp a priori estimation technique for cost versus accuracy remains an elusive goal.

7.1. Periodic problems

To achieve more precise insights into how the formal order of a method affects its accuracy, we consider the model problem $\partial u/\partial t+\partial u/\partial x=0$ on $[-1,1]$, integrated in time from 0 to 2 (the time it takes the analytical solution $u(x,t)$ to move once across the period). The data in Figure 4.1-3 can be recast into Figure 7.1-1 (for details, see Fornberg 1987). The figure should be interpreted as follows.

> Using an accurate time integrator, Fourier modes in the numerical solution of $\partial u/\partial t+\partial u/\partial x=0$ will develop phase errors but not amplitude errors. If a phase error is π, then that mode will have the wrong sign and will not add any accuracy to a Fourier expansion. We here (somewhat arbitrarily) consider a mode to be "accurate" if its phase error is less than $\pi/4$.
>
> An FD2 method with $N_G=500$ (i.e. 500 gridpoints in the spatial direction) is seen to give the same accuracy (have the same horizontal position in the figure) as an FD4 method with $N_G\approx 160$ and a PS method with $N_G\approx 32$. The numbers on the axes represent the following.
>
> Horizontally: the number of modes that remain accurate at the end of the integration. In this example, the horizontal axis position was approximately 16, corresponding to a requirement to keep modes up to $\sin(16\pi x)$ and $\cos(16\pi x)$ accurate.
>
> Vertically: the number of gridpoints needed per wavelength.

Example. Compare FD versus Fourier–PS for the calculation of elastic wave propagation in a 2-D medium.

The governing equations for the test case shown in Figure 7.1-2(a) and Figure 7.1-2(b) are those in Figure 5.3-1 (here using a regular grid). A sharp "pressure wave" pulse is sent down through an elastic medium that carries both pressure waves and shear waves with lower velocities near the center of the domain. After focusing and the subsequent development of a cusp-shaped wave front, Figure 7.1-2(b) shows results for FD2, FD4,

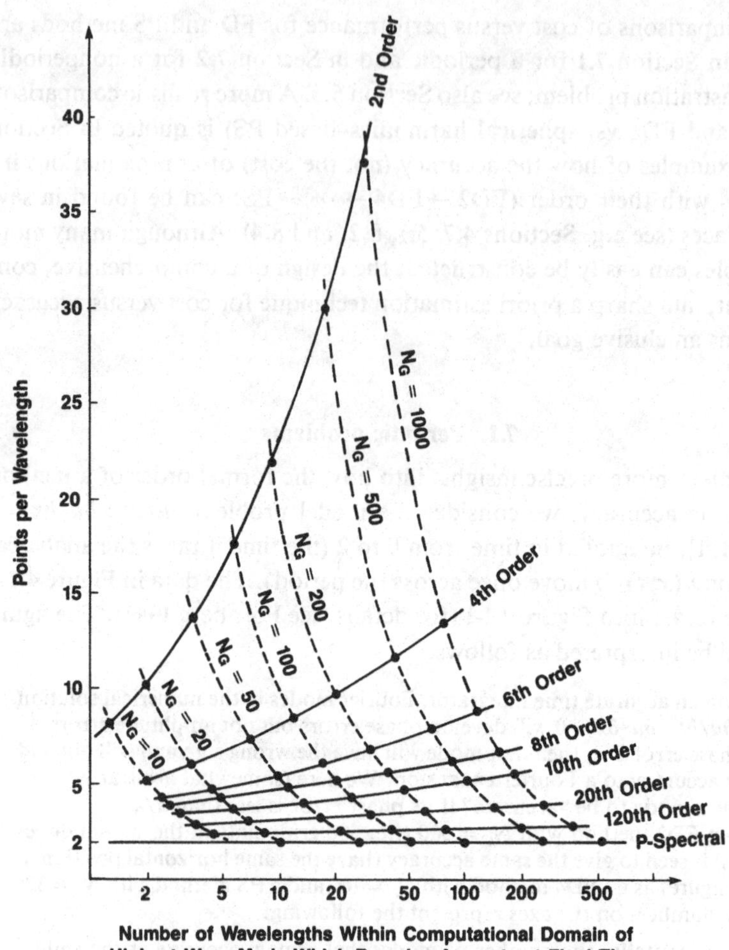

Figure 7.1-1. Relations between grid densities and obtained accuracies when applying different methods to a model problem.

and PS. In the three cases, comparable accuracies are obtained on grids of densities 512×512, 128×128, and 32×32 (resp.) – in quite good agreement with the foregoing discussion. With the PS method implemented by FFTs, the relative computer times scale as 20 : 2 : 1. In three dimensions, the corresponding numbers would become 300 : 8 : 1. For memory requirements, the differences become even larger: 256 : 16 : 1 in two dimensions and 4096 : 64 : 1 in three dimensions.

The accuracy requirement in this example was very moderate. Figure 7.1-1 shows clearly that the relative advantage of PS versus FD methods grows

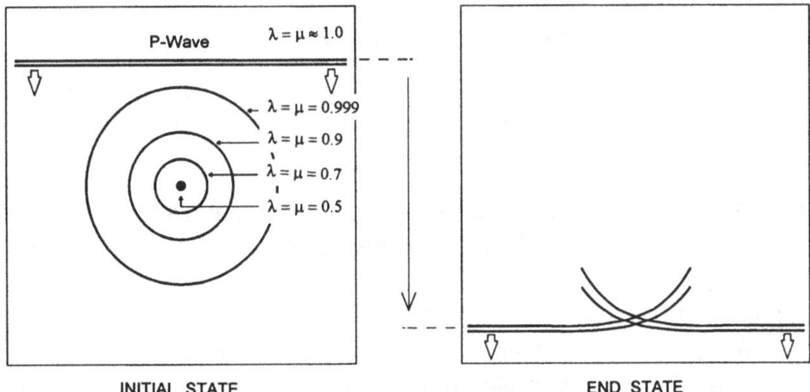

Figure 7.1-2. (a) Contour curves for the variable medium, and schematic illustrations of the initial and end states of the test runs.

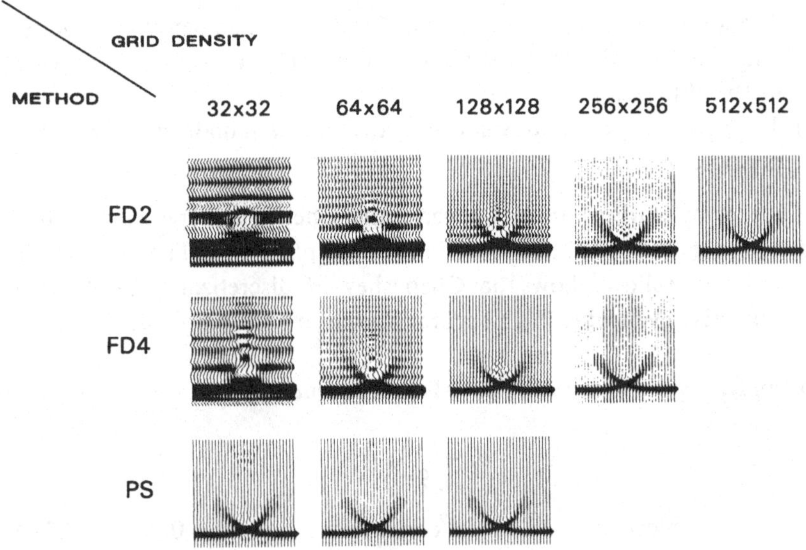

Figure 7.1-2. (b) Numerical results for the test problem (variable f displayed); comparison between different methods and grid densities.

if error tolerances become more stringent. The estimates in the example also illustrate that the advantages with high orders tend to increase with the number of space dimensions.

Hou and Kreiss (1993) note that in one dimension and with near-singular solutions (e.g. with thin internal layers to resolve), FD4 and FD6 methods can match (or exceed) the PS method in efficiency.

7.2. Nonperiodic problems

When passing from periodic to nonperiodic problems, FD and PS methods both require the implementation of boundary conditions. For PS methods, several additional issues arise.

(1) Grids need to be clustered near the boundaries. This can lead to:
  - conditioning and stability problems (especially notable when time stepping);
  - the appearance of spurious EVs, especially for high derivatives (cf. Merryfield and Shizgal 1993 on the KdV equation – in sharp contrast to a very favorable situation for periodic PS methods as described in Section 8.2);
  - the need for preconditioners; and
  - reduced ability to resolve Fourier modes (π vs. 2 points/wavelength are needed).

(2) The formal order of accuracy "only" equals the number of gridpoints, and is not infinite as in the periodic case. (However, the significance of this difference is unclear.)

(3) The performance in nonsmooth cases is less well understood than for periodic problems.

The complications are not severe, and Chebyshev-type PS methods have proven successful in a wide range of applications. The very simple example that follows shows that Chebyshev–PS discretization can be very cost-effective even when high accuracy levels are not required.

Example. Solve numerically the 1-D heat equation

$$u_t = \frac{2}{9\pi^2} u_{xx}$$

$$\text{subject to} \qquad \text{initial condition:} \quad u(x,0) = 0, \qquad (7.2\text{-}1)$$

$$\text{boundary conditions:} \begin{cases} u(0,t) = \sin t, \\ u_x(1,t) = 0, \end{cases}$$

and compare the efficiencies for some different methods in time and space.

Equation (7.2-1) has, for $t > 0$, the analytical solution

$$u(x,t) = \frac{\cos\dfrac{3\pi x}{2}\sinh\dfrac{3\pi(1-x)}{2}\sin t - \sin\dfrac{3\pi x}{2}\cosh\dfrac{3\pi(1-x)}{2}\cos t}{\sinh\dfrac{3\pi}{2}}$$

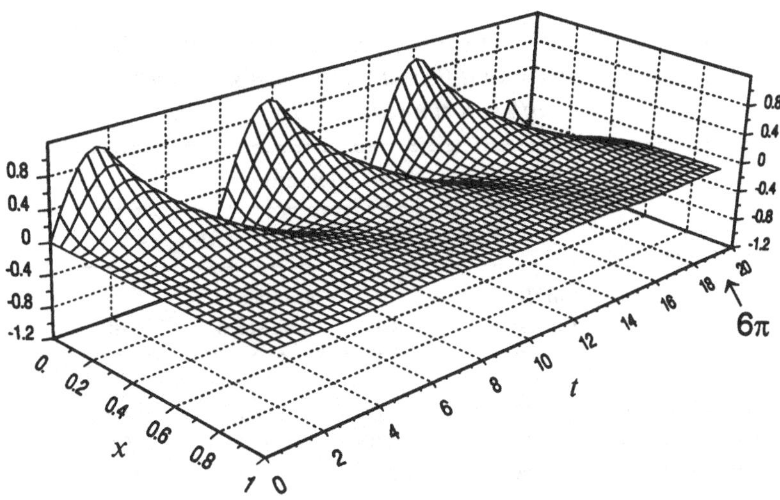

Figure 7.2-1. Analytic solution to the heat equation test problem (equation 7.2-1).

$$+\frac{72}{\pi}\sum_{k=1}^{\infty}\frac{(2k-1)e^{-t(2k-1)^2/18}\sin(k-\frac{1}{2})\pi x}{[9+4(k-2)^2][9+4(k+1)^2]}. \tag{7.2-2}$$

Figure 7.2-1 shows this solution for $0 \le t \le 6\pi$. The top line in equation (7.2-2) describes the long-time asymptotic behavior. The contribution from the infinite sum decays rapidly in time. It is needed for the initial condition $u(x,0)=0$ to be satisfied.

Equation (7.2-1) models temperature variations at different depths underground from day–night or seasonal surface heating–cooling cycles. When using appropriate dimensional units, it transpires that for many soils, the seasonal temperatures are off-phase (warmest in winter, coolest in summer) at depths of a couple of meters.

The FD and Chebyshev–PS DMs for u_{xx} have real, negative eigenvalues. Figure 4.4-2 showed that, for the PS method, some EVs may become quite large in magnitude. Considering the stability domains shown in Appendix G, this suggests a BDF method as a natural choice for the time integration.

Table 7.2-1 and Figure 7.2-2 compare equi-spaced FD2 and FD4 methods with the Chebyshev–PS method for different levels of discretization N when used with BDF and RK time integrators. The symmetry at $x=1$ is implemented as in Example 3 of Section 5.1. In all cases, time steps were chosen as large as possible, subject to maintaining numerical stability and leaving final errors dominated by space (and not time) discretization. The following comments concern implementation of the time integrators.

Table 7.2-1. *Comparisons of costs when solving a 1-D heat equation with different methods in space and different time integrators*

		Time integrators			
		BDF4		RK4	
N	Error	Number of time steps	Cost	Number of time steps	Cost
FD2					
10	.1E−1	25	15,000	60	9,000
20	.3E−2	35	42,000	250	75,000
40	.6E−3	50	120,000	1,000	600,000
FD4					
5	.3E−1	15	7,500	20	2,500
10	.1E−2	50	50,000	90	22,500
20	.5E−4	90	180,000	330	165,000
40	.3E−5	190	760,000	1,300	1,300,000
Chebyshev-PS					
4	.1E−2	40	12,800	35	2,800
5	.1E−3	100	50,000	80	10,000
6	.1E−4	200	144,000	160	28,800
7	.3E−6	400	392,000	290	71,050
8	.1E−7	600	768,000	490	156,800

BDF: BDF1 is known as the *backward Euler* method and, in connection with the heat equation, as Laasonen's method (see Richtmyer and Morton 1967, Sec. 8.2). The main drawbacks with BDF methods involve:

- the difficulty in generating data on previous level(s) in order to get started (assuming $p > 1$); and
- their relatively large cost (one $N \times N$ linear system must be solved per time step).

Because these drawbacks are largely independent of the order p, it makes sense to choose p relatively large. In this example, we use $p = 4$.

The computational costs for any BDF method (in time) together with those for FD2, FD4, and PS (in space) have been estimated at $60N$, $100N$, and $20N^2$ multiplications per time step, respectively. These estimates assume the use of an effective iterative solver (i.e., to obtain representative

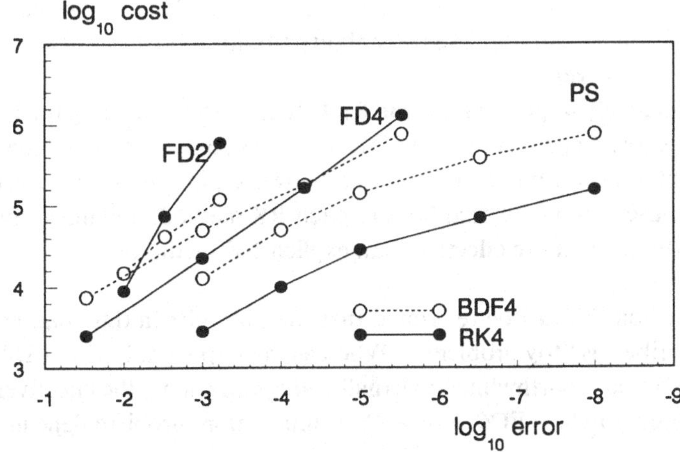

Figure 7.2-2. Cost versus accuracy for equi-spaced FD2 and FD4 and for Chebyshev–PS space discretizations when using the two different time stepping schemes BDF4 and RK4.

comparisons, we do not exploit that – in this special 1-D situation – direct solvers would be faster in the two FD cases).

RK: Especially in the Chebyshev–PS case, it turns out to be essential to implement the Dirichlet boundary condition as described in Appendix G (Section G.1c). The present code implements the BC by using time derivatives of $u(0,t)=\sin(t)$. In the highest-accuracy situation, this reduces the number of time steps needed from 2000 to 490.

The numbers of multiplications per time step have in this case (RK4) been estimated at $15N$, $25N$, and $5N^2$, respectively.

The following observations, from Table 7.2-1 and Figure 7.2-2, regard the effectiveness of the different space and time discretizations.

Space: Especially for RK4, but also for BDF4, cost effectiveness clearly increases with increasing spatial accuracy (FD2 → FD4 → PS).

Time: Due to stability constraints, explicit time stepping methods received a very bad reputation for solving diffusive problems back in the days when nobody considered using anything beyond FD2 in space. Although the stability condition changes from a relatively bad $k \cdot N^2 <$ const. for FD2 to an apparently disastrous $k \cdot N^4 <$ const. for Chebyshev–PS, the change in this example is more than offset by the fact that, for a given accuracy, a much lower value of N suffices. Techniques such as that used in Example 1A of Section 5.1 can improve the stability situation further still (this was not exploited here).

The relatively large extent of the RK4 stability domain along the imaginary axis (see Appendix G) makes the RK4–PS combination possible for many convection-dominated problems. However, in cases of more severe stiffness, Lie (1993) finds that both implicit RK methods and linear multistep methods are more effective than explicit RK methods.

In conclusion, it must be reiterated that the examples in this chapter are best described as "toy problems". Whereas the step from FD2 to FD4 appears to be very worthwhile in virtually all applications, the effectiveness of continuing FD4 → FD6 → ⋯ → PS is much more problem dependent.

8

Applications for spectral methods

Pseudospectral methods originated in the early 1970s out of efforts to improve the accuracy and efficiency for numerical calculations in meteorology and turbulence modeling. PS (and other spectral) methods quickly became well established in these areas. However, other applications have also emerged, such as the solution of different kinds of wave equations. Sections 8.1–8.4 describe very briefly the role of spectral methods in four different areas.

8.1. Turbulence modeling

Even the finest details of turbulence appear to be described by the Navier–Stokes (NS) equations. In their simplest case (2-D incompressible flow), these can be formulated in terms of streamfunction and vorticity.

With u and v denoting the velocity components of fluid particles in the x and y directions, the incompressibility condition $\partial u/\partial x + \partial v/\partial y = 0$ allows a streamfunction ψ to be introduced satisfying $\partial\psi/\partial y = u$, $\partial\psi/\partial x = -v$ (i.e., at each instance, fluid particles move in a direction of constant ψ). The vorticity ω describes the rotational movement of fluid particles: $\omega = \partial v/\partial x - \partial u/\partial y$. In the NS equations

$$\frac{\partial\omega}{\partial t} = \frac{\partial\psi}{\partial x}\cdot\frac{\partial\omega}{\partial y} - \frac{\partial\psi}{\partial y}\cdot\frac{\partial\omega}{\partial x} + \mu\left(\frac{\partial^2\omega}{\partial x^2} + \frac{\partial^2\omega}{\partial y^2}\right), \tag{8.1-1}$$

$$\omega = -\left(\frac{\partial^2\psi}{\partial x^2} + \frac{\partial^2\psi}{\partial y^2}\right), \tag{8.1-2}$$

(8.1-1) expresses that vorticity is convected along streamlines and, at the same time, diffused like the solution to a 2-D heat equation (μ is a measure of the fluid viscosity). When (8.1-1) is time stepped, the streamfunction ψ must be continuously updated through repeated solution of the Poisson equation (8.1-2) (which expresses how the definitions of ψ and ω link these two quantities with each other).

The main computational difficulty is in economically representing a wide enough range of scales - from the full extent of the fluid (e.g. fluids in vessels, oceans, or intergalactic jets) down to the Kolmogorov scale (where molecular effects create a diffusion that firmly dominates over convection). There are two main ways to address this difficulty.

Large eddy simulation (LES): Only the largest scales are modeled directly. The influences of smaller scales are accounted for by *subgrid modeling* - further equations that (it is hoped) model averaged properties of fine-scale turbulence.

Direct numerical simulation (DNS): Here one attempts to resolve all scales, but restricts the calculation to low Reynolds numbers (large values of μ in (8.1-1)).

To reach a more favorable compromise between the conflicting goals of

- a wider range of frequency components that are little affected by viscosity (a larger "inertial range") and
- a stronger damping of the very highest modes (leading to improved stability and reduced aliasing),

the concept of *hyperviscosity* is often employed. In case of equation (8.1-1), the diffusive term $\mu(\partial^2/\partial x^2+\partial^2/\partial y^2)\omega$ may be replaced by a nonphysical hyperdiffusive term such as $-\mu_1(\partial^2/\partial x^2+\partial^2/\partial y^2)^2\omega$ (or, more drastically, all Fourier modes above some threshold may be removed at each time step). Although this can speed up computations dramatically, it can also be a source of errors (Browning and Kreiss 1989). Yao, Dritschel, and Zabusky (1994) observe error trends that, in their words, "bring into question the results of many 2D turbulence simulations."

In the first PS application (of any kind), Fox and Orszag (1973) time stepped (8.1-1) and (8.1-2) as outlined here. The sequence in Figure 8.1-1 (from Fornberg 1977) shows how a very irregular initial vorticity field organizes itself into fewer and larger vortices. This process - predicted from different approaches by Onsager (1949) and Batchelor (1953) - is specific to two dimensions. It has relevance to flows in layers where the vertical extent is much less than the horizontal ones.

Normally occurring turbulence is a genuinely 3-D phenomenon of vast practical implications. Large-scale motions generate features on finer scales, which then evolve in chaotic manners. Both qualitative and quantitative modeling of turbulence pose outstanding scientific and engineering challenges.

When the first vector supercomputers (Cray 1 and CDC Cyber 205) became available toward the end of the 1970s, full 3-D simulations of turbulence could be attempted. Brachet et al. (1983) exploited assumed

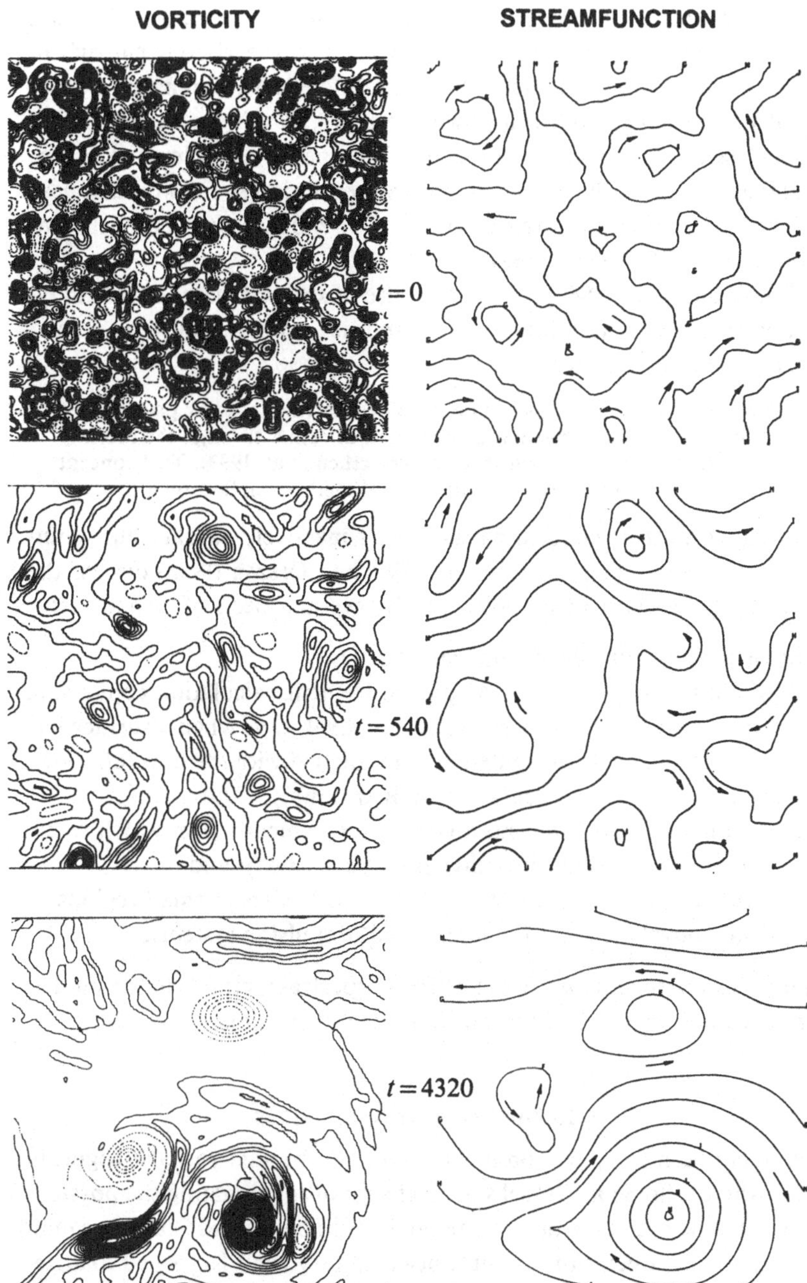

Figure 8.1-1. Two-dimensional turbulence: a randomly generated initial vorticity field organizing itself into large vortices (results from a PS calculation on a 64^2 grid, with t denoting the number of time steps).

symmetries to obtain numerical evidence of an *inertial range*. This is a range of frequencies (between large eddies and the viscous cut-off) in which the distribution of energy between different wavenumbers is theoretically expected to follow a power law, $E(t,k)=k^{-m}$, where k is the magnitude of the wavenumber vector. The theory of Kolmogorov suggests that $m=5/3$. The highest numerical resolution so far would appear to be a 512^3 calculation carried out on a CM-200 massively parallel computer system (Chen et al. 1993).

Orszag and Kells (1980) initiated the use of spectral methods for studying the structures of nonlinear 3-D instabilities at walls, which can lead to a transition to turbulence.

The onset of wall instabilities appears to be related to a linear process best described not through eigenmode analysis but rather through the recently introduced concept of *pseudospectra* (Trefethen et al. 1993). This concept figured in the stability discussion of spectral methods in Section 4.5.

In the context of advocating the use of massively parallel supercomputers in fluids calculations, Karniadakis and Orszag (1993) discuss the use of PS (or spectral) methods to study several aspects of turbulence:

- homogeneous turbulence (no wall effects);
- drag reduction by riblets – a spectral element calculation confirming experimentally observed drag reduction by 4–12% when a surface has a fine, sharp-edged corrugated pattern aligned with the flow (although a small drag increase occurs for laminar flows);
- supersonic compressible homogeneous turbulence (for which 512^3-resolution calculations also have been performed); and
- supersonic reacting shear layers – detailed studies in small regions to evaluate novel concepts for high-speed propulsion systems.

Numerous references on the use of PS or spectral methods in the modeling of turbulence can be found in Canuto et al. (1988).

8.2. Nonlinear wave equations

The scientific history of nonlinear waves can be said to have begun in 1834, when John Scott Russel was in charge of investigating the possibility of introducing steam navigation on the Edinburgh and Glasgow canal. During a barge operation, a single peak solitary wave happened to arise. He followed it on horseback for over a mile (past some bends), recognizing immediately that such a *wave of translation* (no troughs in front of or behind the single peak) represented something novel and fundamentally different from the wavetrains that are seen much more commonly. In subsequent wavetank experiments with solitary waves, he noted again their stability and also the tendency of pulses to break up into such solitary waves.

In 1895, Korteweg and de Vries derived

$$u_t + uu_x + u_{xxx} = 0 \tag{8.2-1}$$

as a model equation for waves on shallow water. This so-called KdV equation admits the solitary wave solution

$$u(x,t) = 3\alpha^2 \operatorname{sech}^2 \tfrac{1}{2}(\alpha x - \alpha^3 t), \tag{8.2-2}$$

where the parameter α determines both the amplitude and the speed.

Two numerical studies offered the first glimpses of the full richness of nonlinear wave phenomena. Fermi, Pasta, and Ulam (1955) studied a discrete nonlinear equation that modeled vibrations in a lattice, and noted curious recurrences in the time evolution (rather than an expected trend away from coherent features in the solutions). Zabusky and Kruskal (1965) noted a connection between this problem and the KdV equation (8.2-1). They applied the leapfrog FD2 scheme

$$\begin{aligned} u_m^{n+1} = u_m^{n-1} &- \frac{k}{3h}(u_{m+1}^n + u_m^n + u_{m-1}^n)(u_{m+1}^n - u_{m-1}^n) \\ &- \frac{k}{h^3}(u_{m+2}^n - 2u_{m+1}^n + 2u_{m-1}^n - u_{m-2}^n) \end{aligned} \tag{8.2-3}$$

to (8.2-1) and observed that the recurrence was a consequence of different solitary waves (8.2-2) re-emerging entirely unchanged (apart from sideways shifts) following nonlinear interactions. (See Aoyagi and Abe 1989 regarding a weak nonlinear instability in (8.2-3).) Gardner, Greene, Kruskal, and Miura (1967) found that (8.2-1) (assuming that solutions decay sufficiently fast at infinity) could be linearized through a technique known as inverse scattering. The solitary wave interactions were proven to be "clean", and such solitary waves becamed known as *solitons*. The resulting equation is described as being integrable.

The interaction of N unequal solitary waves (8.2-2) with parameters $\alpha_1, \alpha_2, \ldots, \alpha_N$ can be written in closed form:

$$u(x,t) = 12\frac{\partial^2 \ln|D|}{\partial x^2}$$

where $|D|$ is the determinant with elements

$$D_{mn} = \delta_{mn} + \frac{2}{\alpha_m + \alpha_n} e^{-\alpha_m(x - x_m) + \alpha_m^2 t}, \quad \delta_{mn} = \begin{cases} 1 & \text{if } m = n, \\ 0 & \text{otherwise.} \end{cases}$$

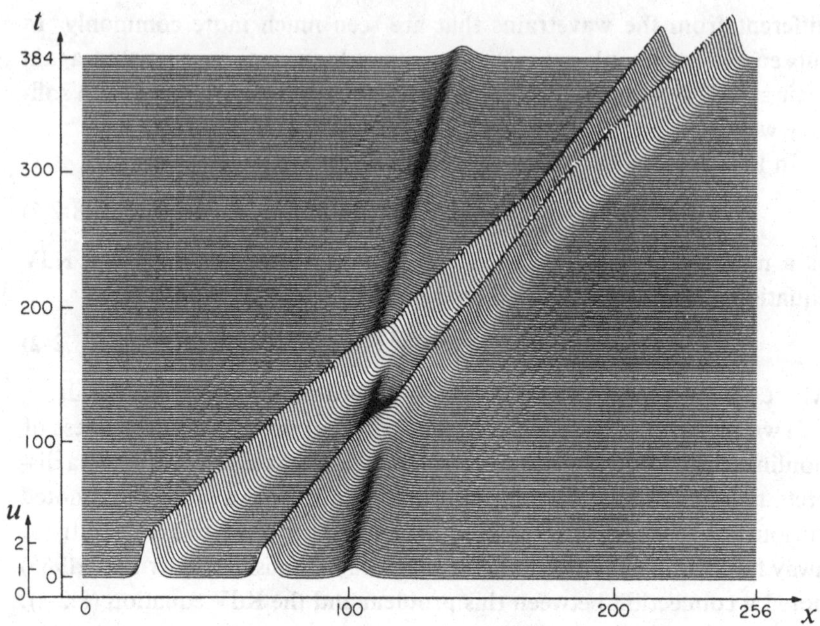

Figure 8.2-1. Interactions of three solitons for the KdV equation $u_t + uu_x + u_{xxx} = 0$ (as calculated with a Fourier–PS method using a space step of $h = 1$; the barely visible ripples are numerical errors).

Figure 8.2-1 shows the interaction of three solitary waves for (8.2-1), as computed by a periodic PS method (Fornberg and Whitham 1978; the picture could equally well have been obtained from the preceding closed-form expression).

Following the discovery of recurrences, solitons, and exact solution methods, research efforts in the area of nonlinear waves expanded rapidly. Many other equations were also found to possess these remarkable properties, including such 1-D continuum equations as:

$$u_t + 3u^2 u_x + u_{xxx} = 0 \quad \text{(modified KdV equation)}, \tag{8.2-4}$$

$$iu_t + u_{xx} + 2|u|^2 u = 0 \quad \text{(cubic nonlinear Schrödinger equation)}, \tag{8.2-5}$$

$$u_{tt} - u_{xx} + \sin u = 0 \quad \text{(sine–Gordon equation)}, \tag{8.2-6}$$

and

$$u_t + 6uu_x + \frac{1}{\pi}\,\mathrm{PV}\int_{-\infty}^{\infty} \frac{u_{\xi\xi}(\xi)}{\xi - x}\,d\xi = 0 \quad \text{(Benjamin–Ono equation)},$$

where PV denotes a principal-value integral. In two dimensions we have

$$(u_t+6uu_x+u_{xxx})_x \pm 3u_{yy}=0 \quad \text{(Kadomtsev-Petviashvili equation)}$$

and furthermore a few difference schemes – either discrete in space only:

$$i\frac{du_m}{dt}+\frac{1}{h^2}(u_{m+1}-2u_m+u_{m-1})+|u_m|^2(u_{m+1}+u_{m-1})=0; \qquad (8.2\text{-}7)$$

or discrete in both space and time:

$$\left(1+\frac{h^2}{4}\right)\tan\frac{u_m^{n+1}+u_m^{n-1}}{4}-\left(1-\frac{h^2}{4}\right)\tan\frac{u_{m+1}^n+u_{m-1}^n}{4}=0. \qquad (8.2\text{-}8)$$

The two schemes (8.2-7) and (8.2-8) approximate to second-order (8.2-5) and (8.2-6), respectively (Ablowitz and Ladik 1976, Hirota 1977).

As can be seen from Figure 8.2.2(a), the FD2 approximation (8.2-7) is very suitable for numerical work. If the nonlinear term is instead approximated by $2|u_m|^2u_m$ (equation (8.2-14), to be discussed shortly), the numerical solution falls apart rapidly – Figure 8.2.2(b). As part (c) shows, using a PS approximation in space for u_{xx} (together with $2|u_m|^2u_m$ for the nonlinear term) cleans up the noise in part (b), but produces a quite different solution to the one in part (a). The analytic solution is unstable, and the tiniest perturbations cause large quantitative differences. An integrable scheme may show a more consistent long-time solution pattern. However, for stable solutions to nonlinear wave equations, a PS scheme is generally much more accurate than an integrable FD2 scheme.

Numerical experiments with (8.2-8) show that an unlimited number of conservation laws and exact soliton interactions need not guarantee even that solutions remain smooth (Ablowitz, Herbst, and Schober 1995b).

Nonlinear integrable wave equations arise in a large number of applications, such as plasma physics, anharmonic crystals, bubble–liquid mixtures, nonlinear optics, and so forth. General references on the analytic theory for this class of equations include Whitham (1974) and Ablowitz and Segur (1981).

Numerical methods

The two main classes of *space* discretizations for nonlinear wave equations are FD and spectral methods.

FD methods: Second- or higher-order FD approximations are applied to the space derivatives in the governing equation. Performance is often influenced by variations that would be of little consequence in most other contexts, as with the nonstandard approximations of the nonlinear terms in (8.2-3) and (8.2-7) (the former corresponds to the case of $\theta=\frac{2}{3}$ discussed in Section 4.6).

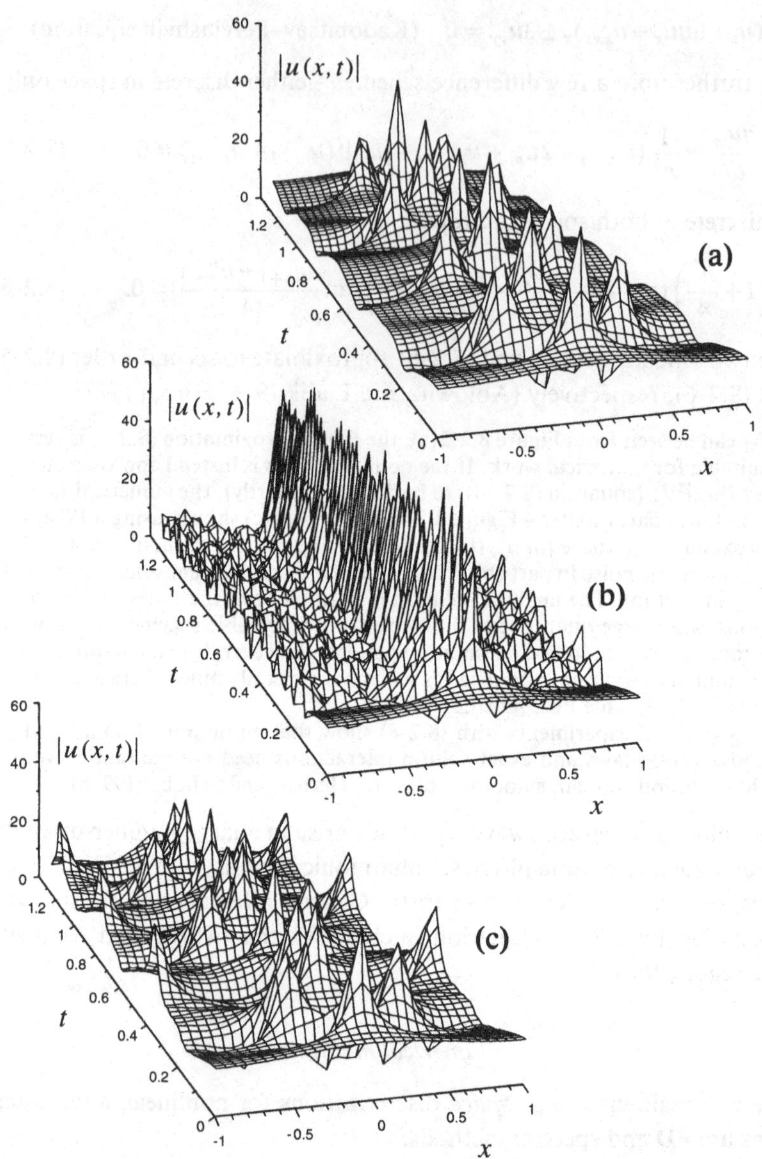

Figure 8.2-2. Recurrences for the cubic nonlinear Schrödinger equation $iu_t + u_{xx} + 2|u|^2 u = 0$ with $u(x, 0) = \pi\sqrt{2}(1 + 0.1\cos\pi x)$ calculated using different spatial approximations ($N = 32$; accurate time integration). **(a)** Integrable scheme $i(\partial u_m/\partial t) + (1/h^2)(u_{m+1} - 2u_m + u_{m-1}) + |u_m|^2(u_{m+1} + u_{m-1}) = 0$. **(b)** Standard FD2 scheme as in (a), but with nonlinear term approximated as $2|u_m|^2 u_m$. **(c)** Standard PS scheme, with nonlinear term approximated as in (b).

Spectral methods: Fourier–PS methods are particularly easy to implement. They are often suitable because the key phenomena frequently

- concern exact (or near-exact) properties – very high computational accuracy is called for;
- involve long-time behaviors of solutions; and
- do not depend on boundary effects – spatially periodic formulations can be used. (Periodic boundary conditions may also be physically relevant – e.g., Stokes waves on deep water, which are known to be unstable.)

Many numerical approaches are available for efficient *time* integration of nonlinear wave equations. The MOL approach is (as usual) very broadly applicable. Following an example of MOL, we note that special features of differential equations often can be exploited in the design of very effective specialized schemes. The three methods that are presented for the KdV equation (8.2-1) all depend on the fact that the highest derivative term enters without a variable coefficient. The last of the time stepping methods described (symplectic integrators) has received much recent attention in other contexts. However, its significance for nonlinear wave calculations is at present unclear.

Method of lines (MOL): Very effective Fourier–PS methods can be obtained for most nonlinear wave equations by simply combining PS spatial differentiation with an ODE solver for the time integration. A recent example of this approach (using a variable-step and variable-order Adams–Bashforth/Adams–Moulton time integrator) is the study by Rosenau and Hyman (1993). They generalize (8.2-1) and (8.2-4) to

$$u_t + (u^m)_x + (u^n)_{xxx} = 0, \quad m > 0,\ n = 2 \text{ or } 3,$$

and note that, with the nonlinear dispersion, solitary waves can have compact support; for example,

$$u(x,t) = \begin{cases} \dfrac{4c}{3}\cos^2\!\left(\dfrac{x-ct}{4}\right) & \text{if } |x-ct| \le 2\pi, \\ 0 & \text{if } |x-ct| > 2\pi, \end{cases}$$

in the case of $m = n = 2$. These solitary waves were found to have many soliton-type features, and hence they were named *compactons*.

Split-step: Tappert (1974) proposes a split-step Fourier scheme for the KdV equation (8.2-1). The general idea of fractional steps was described in Section 4.5. In this special case, we note that the nonlinear equation $u_t + 2uu_x = 0$ can be advanced in time by an FD method and the linear equation $u_t + 2u_{xxx} = 0$ by modifying phase angles of Fourier coefficients. The first step may impose a mild stability condition; the second step is unconditionally stable.

Although most time splitting methods are limited to second-order accuracy in time, Yoshida (1990) notes that split-step symplectic methods (to be described shortly) can be accurate to arbitrary orders in time (and in space).

Leapfrog–Fourier–PS: Fornberg and Whitham (1978) note that immediate MOL leapfrog–Fourier–PS approximations of the KdV equation can be improved on by a simple modification. Let F denote the discrete Fourier transform from physical x space to Fourier ω space. On the periodic interval $[-1, 1]$, the Fourier modes are $e^{i\pi\omega x}$, $\omega = 0, \pm 1, \pm 2, \ldots, \pm N/2$. The straightforward PS approximation with leapfrog time discretization becomes

$$u(x, t+k) - u(x, t-k) + 2ik\pi u F^{-1}\{\omega F u\} - 2ik\pi^3 F^{-1}\{\omega^3 F u\} = 0. \quad (8.2\text{-}9)$$

We modify the last term in the left-hand side to obtain

$$u(x, t+k) - u(x, t-k) + 2ik\pi u F^{-1}\{\omega F u\} - 2iF^{-1}\{\sin(k\pi^3\omega^3) F u\} = 0. \quad (8.2\text{-}10)$$

For low wavenumbers ω, the difference between (8.2-9) and (8.2-10) is only $O(k^3)$, that is, on the same level as the error incurred in the time stepping. For high wavenumbers ω, the term u_{xxx} dominates u_x and the KdV equation becomes essentially $u_t + u_{xxx} = 0$. Although the time step looks like a conventional leapfrog, the scheme (8.2-10), without the nonlinear term, is exact for this linear equation (for all k and ω).

Equation (8.2-10) is more accurate than (8.2-9) for high ω; moreover, linearized von Neumann stability analysis shows the stability condition to have been relaxed nearly fivefold – from $k/h^3 < 1/\pi^3$ to $k/h^3 < 3/(2\pi^2)$. This latter condition is not a serious restriction. In view of the second-order accuracy in time and better than sixth-order accuracy in space, the time steps are quite likely to be more limited by accuracy than by stability. The computationally most costly part is the FFTs, of which three are needed for each time step.

Explicit spectral: Chan and Kerkhoven (1985) note that writing the KdV equation (8.2-1) in the form $u_t + \frac{1}{2}(u^2)_x + u_{xxx} = 0$ allows for a particularly effective spectral implementation. Let superscripts denote time levels and the symbol ˆ above a variable signify its Fourier coefficient (for wavenumber ω). Assuming that u and $\hat{u}$ are known for time levels $n-1$ and n, these quantities for level $n+1$ are obtained through

$$\hat{u}^{n+1} = \hat{u}^{n-1} - i\omega k F((u^n)^2) + ik\omega^3(\hat{u}^{n+1} + \hat{u}^{n-1});$$

$$u^{n+1} = F^{-1}\hat{u}^{n+1}.$$

Strengths of this method include:

- reduced computation – each time step requires only two FFTs; and
- explicitness and unconditional stability.

Symplectic integrators: Systems of ODEs are called *Hamiltonian* if they can be cast in the form

$$\frac{\partial p_n}{\partial t} = -\frac{\partial H(p,q)}{\partial q_n}, \quad \frac{\partial q_n}{\partial t} = \frac{\partial H(p,q)}{\partial p_n}, \tag{8.2-11}$$

where p, q are vectors with elements $p_n(t), q_n(t)$, respectively, $n = 1, 2, \ldots, N$, and H is some function of p and q.

Equation (8.2-11) implies immediately that $dH/dt = 0$; i.e., $H(p,q)$ is constant in time. Conversely, many nondissipative physical systems can be written in Hamiltonian form (with energy serving as the Hamiltonian function).

Even in unstable and chaotic solutions to Hamiltonian systems, the functions p and q will evolve in a *symplectic* manner.

If $N = 1$, the area of any arbitrary domain in a (p, q) plane is time invariant. If $N > 1$, the same holds for the sum of the N projections onto the (p_n, q_n) planes $(n = 1, 2, \ldots, N)$.

Symplectic integrators (which include some implicit RK methods) preserve exactly this same quantity, even for arbitrarily large time steps. Especially for low values of N, the combination of a Hamiltonian formulation and a symplectic time integrator has proven very successful for a wide class of dynamical systems. For example, in many oscillator problems, the error grows linearly in time for symplectic integrators, but quadratically for dissipative ones (such as explicit RK methods). The topic is surveyed in Sanz-Serna and Calvo (1994); McLachlan (1994) describes how symplectic integrators offer a numerical approach to the classical problem of the long-term stability of planetary orbits in the solar system.

It turns out that many of the standard FD and PS approximations to common nonlinear wave equations happen to be of Hamiltonian form.

For any Hamiltonian nonlinear wave equation, one can always find a finite dimensional approximation that also is Hamiltonian. This approximation can then be advanced in time through a symplectic integrator. Still, there appear to be no entirely straightforward methods available to answer any of the following questions:

- Can a given nonlinear wave equation be cast in a Hamiltonian form?
- Is a given discrete approximation of Hamiltonian form?
- Is there a best symplectic discretization (and what should be meant by "best")?

Example. State a Hamiltonian H for the nonlinear Schrödinger (NLS) equation (8.2-5) and construct from this an FD2 scheme that is of the form (8.2-11).

Table 8.2-1 lists the first four of the infinitely many conservation laws (Hamiltonians) available for the NLS equation. By forming *variational derivatives* of these (a continuous version of the discrete process we will employ in this example), a particular PDE can be recovered from each law. The third conservation law turns out to return precisely the same NLS equation that we started with.

We consider the following (one-sided) FD1 approximation to the Hamiltonian

$$H = i\sum_n \left(\frac{|u_{n+1}-u_n|^2}{h^2} - |u_n|^4\right). \tag{8.2-12}$$

(The factor i, the imaginary unit, turns out to be necessary to obtain the NLS equation.) Instead of immediately trying to solve for $u_n(t)$, $n=1,2,\ldots,N$, we introduce new variables and rewrite the Hamiltonian as

$$H = -i\sum_n \left[\frac{(q_{n+1}-q_n)(p_{n+1}-p_n)}{h^2} + p_n^2 q_n^2\right], \tag{8.2-13}$$

noting that if $p_n = -\bar{u}_n$ and $q_n = u_n$, $n=1,2,\ldots,N$, the Hamiltonian (8.2-12) is recovered. Differentiation of (8.2-13) gives

Table 8.2-1. *The first four conservation laws for the NLS equation and their variational derivatives*

Initial equation	Conserved quantities $H(u,\bar{u})$	Equation recovered from each H as the variational derivative
$iu_t+u_{xx}+2\|u\|^2u=0$	$i\int\|u\|^2dx$	$u_t+iu=0$
	$\int u\bar{u}_x\,dx$	$u_t-u_x=0$
	$i\int(\|u_x\|^2-\|u\|^4)\,dx$	$iu_t+u_{xx}+2\|u\|^2u=0$
	$\int(u_x\bar{u}_{xx}-4u^2\bar{u}\bar{u}_x-\bar{u}^2uu_x)\,dx$	$u_t+6\|u\|^2u_x+u_{xxx}=0$
	$\vdots$	$\vdots$

$$\frac{\partial H}{\partial p_n} = \frac{i}{h^2}(q_{n+1} - 2q_n + q_{n-1}) - 2ip_n q_n^2,$$

$$\frac{\partial H}{\partial q_n} = \frac{i}{h^2}(p_{n+1} - 2p_n + p_{n-1}) - 2iq_n p_n^2,$$

with slight modifications for $n = 1$ and $n = N$.

With these values for $\partial H/\partial p_n$ and $\partial H/\partial q_n$, equation (8.2-11) describes a time evolution of $p_n(t)$ and $q_n(t)$ that leaves $H(p, q)$ invariant (as required). Choosing $p_n = -\bar{u}_n$ and $q_n = u_n$, $n = 1, 2, \ldots, N$, we obtain two formulations of the following relation (as it stands, and its complex conjugate):

$$\frac{\partial u_n}{\partial t} = \frac{i}{h^2}(u_{n+1} - 2u_n + u_{n-1}) + 2i|u_n|^2 u_n. \qquad (8.2\text{-}14)$$

This "routine" FD2 approximation (8.2-14) to the NLS equation (8.2-5) could of course have been written down immediately without any of the preceding algebra. The point of this example was only to establish that (8.2-14) is indeed of Hamiltonian form, and hence could possibly benefit from the use of a symplectic time integrator. (However, as we saw in Figure 8.2-2(b), the scheme (8.2-14) is not well suited for studying recurrences of (8.2-5).)

As noted earlier, the importance of symplectic integrators in the field of nonlinear wave equations is at present unclear. Ablowitz et al. (1995a,c) find that both high-order spatial approximations and large values of N seem to reduce or eliminate whatever advantages such integrators may have in other contexts.

Comparison of accuracy – FD versus Fourier-PS

Numerous comparisons have been performed between different numerical methods for the KdV (and similar) equations. None of the FD (and FE) schemes compared in Vliegenthart (1971) come close to the performance of the spectral schemes that were developed shortly afterward. Taha and Ablowitz (1984) find that one inverse scattering-based scheme runs their tests slightly faster than the Fornberg–Whitham scheme (and much faster than the Tappert split-step method). Nouri and Sloan (1989) find the Chan–Kerkhoven scheme to be the fastest one in their tests. They also describe an efficient split-step characteristic scheme (for the KdV equation).

The following example (from Fornberg and Whitham 1978) shows that nonlinear wavetrain instabilities offer an opportunity to compare how errors due to different spatial discretization methods accumulate over time, without the need to numerically carry out any actual time integration.

Example. Compare the growth rates of the unstable sidebands for FD2, FD4, ... and Fourier–PS approximations to uniform wavetrain solutions for the modified KdV equation (8.2-4).

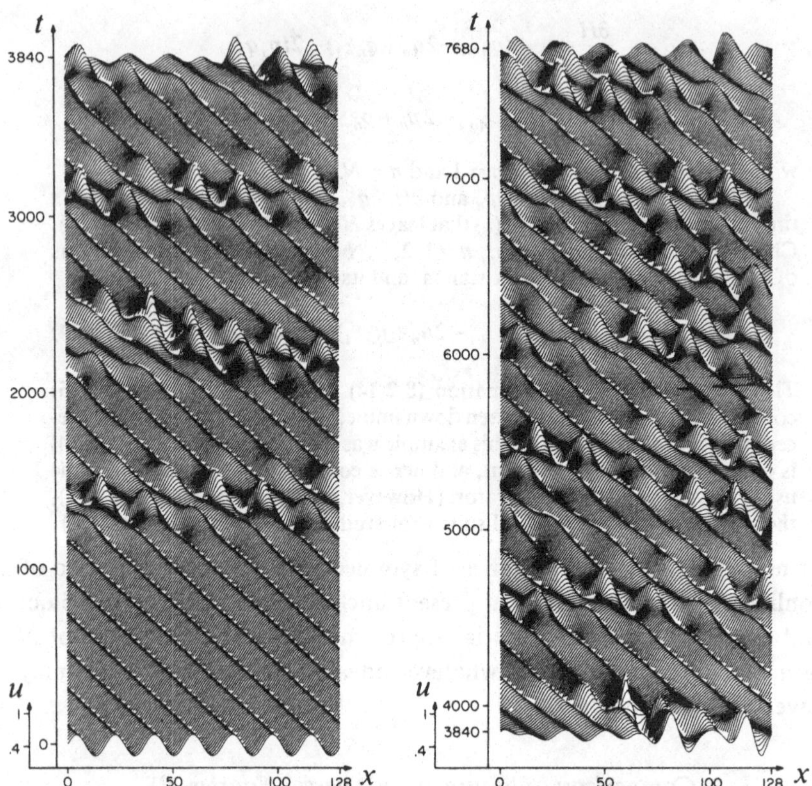

Figure 8.2-3. Fourier–PS solution of the modified KdV equation $u_t + 3u^2u_x + u_{xxx} = 0$, with a pure sinusoidal initial condition $u(x,0) = 0.2\sin(7\pi/64)x$.

Figure 8.2-3 shows how a solution with the initial condition $u(x,0) = 0.2\sin(7\pi/64)x$ evolves over time in a Fourier-PS calculation. Like Figure 8.2-2(c), this displays a locally very accurate approximation to an unstable situation. Even the smallest perturbations will alter quantitative (but not qualitative) properties of the solution. The main mode – seven oscillations in $[0,128]$ – is unstable to exponentially growing sidebands. The sideband that grows the fastest is in this case a combination of the modes 9 and 5 (when denoting the initial condition as mode 7). Figure 8.2-4 shows the quantity amplitude^2/wavenumber for the different Fourier modes as the solution evolves in time. Around time $t = 1400$, the sidebands are seen to first grow exponentially and then decay again.

For the numerical comparison of different methods, we choose slightly different initial conditions – the values at $t = 0$ of the translating wave-train solution

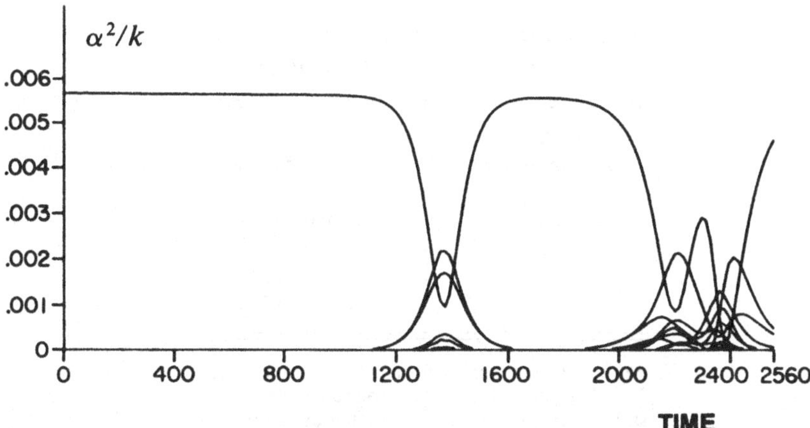

Figure 8.2-4. Time evolution of different Fourier modes in the solution shown in Figure 8.2-3 (amplitude² divided by wavenumber).

$$u(x,t)=u_0(x-ct)=0.3\cos\left[\frac{7\pi}{64}(x-ct)\right]+0.00769624\cos\left[3\frac{7\pi}{64}(x-ct)\right]$$

$$+0.00019263\cos\left[5\frac{7\pi}{64}(x-ct)\right]+0.00000482\cos\left[7\frac{7\pi}{64}(x-ct)\right]$$

$$+0.00000012\cos\left[9\frac{7\pi}{64}(x-ct)\right]+\cdots,\quad c=-0.04874619. \qquad (8.2\text{-}15)$$

Analysis of the unstable modes becomes simpler if we translate to a frame of reference that also moves with speed c. Equation (8.2-4) then becomes

$$u_t+(3u^2-c)u_x+u_{xxx}=0.$$

Unstable modes are found by substituting

$$u(x,t)=u_0(x)+e^{\alpha t}v_0(x)$$

into this equation, where v_0 is a small perturbation. Ignoring second-order quantities in v_0 leads to an eigenvalue problem with the growth rates as the eigenvalues α listed in Table 8.2-2.

For any numerical scheme that discretizes the spatial variable, we can proceed in the same way and obtain corresponding numerical growth rates by solving a matrix eigenvalue problem. Figure 8.2-5 shows how the largest error in any of the four growth rates depends on the method used and the number of gridpoints.

Table 8.2-2. *Growth rates for different sidebands in the model problem*

Dominant sideband modes	Growth rate α
6, 8	0.0101916188
5, 9	0.0184643383
4, 10	0.0220767943
3, 11	0.0130214545

Notes: The main mode is denoted as mode 7 (i.e., 7 oscillations in the period). The four different unstable perturbations are dominated by the pair of modes listed (but contain also higher harmonics).

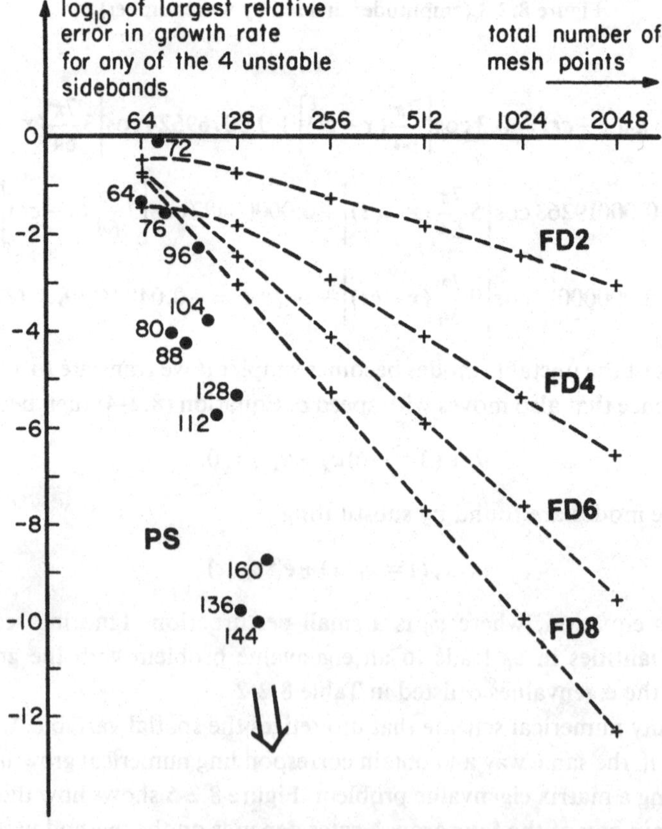

Figure 8.2-5. Accuracy of sideband growth rates for different methods and grid densities.

The irregularities for the PS method arise because high harmonics in (8.2-15) for some N on the discrete grid can be misinterpreted as low modes and so "contaminate" some of the sideband modes (an unusually noticeable instance of aliasing errors).

As N increases, the PS method is seen to converge much faster than FD methods. For the high accuracies that are often required for nonlinear wave problems (in the range of 10^{-8} to 10^{-14}), low-order FD methods require enormous grid densities. Explicit time stepping can be quite acceptable for the PS method (even when subject to a k/h^3-type stability condition), but is entirely unacceptable for low-order FD methods.

8.3. Weather prediction

Meteorological data were first collected in a systematic manner during the first half of the 18th century, following the discovery of the barometer in 1643 and the first temperature scale in 1714. The first map of surface pressures was produced in 1820 (using nearly 50-year-old data). Shortly after the invention of the telegraph, the first current weather maps were displayed at the 1851 London World's Fair. Skilled meteorologists assessing up-to-date measurements (and past histories) remained the dominant prediction approach well into the 1950s.

V. Bjerkness proposed in 1904 that, in principle, it should be possible to predict the weather from the laws of fluid mechanics. A pioneering – but premature – attempt was made by L. F. Richardson while working as an ambulance driver on the Western Front during the First World War; it is described in his book of 1922. For future prediction work, he envisaged a great dome with tens of thousands of human calculators arranged according to their assigned domain on the surface of the earth. Data was to be passed between different areas and also communicated to the center, from where the whole operation was to be controlled in a manner somewhat reminiscent of how an orchestra is conducted.

Several of the difficulties that lay ahead concerned instabilities, both physical and numerical.

Intrinsic instability of the weather

Tiny perturbations can grow exponentially. Although various conservation laws impose global restrictions on the weather, the "butterfly effect" suggests that a single butterfly, flapping its wings, can affect the path that hurricanes take a year later.

Short-term local forecasts and long-term climatic (time-averaged) forecasts are possible, but accurate long-term local forecasts are impossible.

Numerical linear instabilities

These were almost spotted by Richardson in 1910 (see Section 4.5), and they caused his early effort to fail quite spectacularly.

Some of these instabilities can be avoided by a good choice of FD stencil. However, a fundamental problem is that the equations of compressible flow also describe sound and gravity waves. Although these waves are not of meteorological interest, their high velocities can still lead to severe stability restrictions for explicit schemes.

Numerical nonlinear instabilities

As was observed first in meteorological calculations (Phillips 1959), numerical schemes can sometimes generate explosively growing (unphysical) spikes even when linear stability conditions are satisfied. This phenomenon was discussed in Section 4.6.

Many factors contributed to a breakthrough in numerical weather forecasting following the Second World War. The work by John von Neumann was critical for the last three of the following factors:

- improved observational network;
- improved understanding of atmospheric dynamics;
- design and construction of electronic computers;
- recognition of numerical weather prediction as an ideal challenge for such computers - and assembling a team of meteorologists for the task; and
- clarification of the issue of linear instabilities (see Section 4.5).

The first successful experimental code was operational in 1950. The first computed forecasts were issued in 1953 (in Sweden), and from 1955 on by the U.S. Weather Service. These codes used FD methods, which were to remain the dominant numerical approach for nearly 30 years.

Spectral methods for weather forecasting can historically be traced back to work in the Soviet Union during the mid-1940s. Proposals were made in the mid-1950s to use spherical harmonics in the horizontal directions together with FD approximations vertically. With Galerkin's method, linear terms were quite straightforward to implement. However, as noted in Section 2.2, nonlinear terms required costly convolutions. The two key events that made spectral methods practical were

- discovery of the FFT algorithm (Cooley and Tukey 1965); and
- the idea of swapping between transform and grid representations in order to simplify treatment of nonlinear terms (Eliasen, Machenhauer, and Rasmussen 1970 and, independently, Orszag 1970).

The grids were typically equi-spaced in longitude (to allow FFTs) and Gaussian in latitude (to allow Gaussian quadrature-type Legendre transforms – cf. the structure of spherical harmonic functions, Section 6.1). The availability of grid data at each time step simplified the implementation of local phenomena (such as precipitation). However, a grid would need to have more points than there are parameters in a spherical harmonics expansion, making conversions between grid and spectral data nontrivial.

The first national weather services to use spectral codes for routine forecasts were Australia and Canada in 1976, followed by the United States (1980), France (1982), and Japan and the European Center for Medium Range Weather Forecast (ECMWF) (1983).

The global spectral codes have largely been based on spherical harmonics and Galerkin approximations. This approach is described in Jarraud and Baede (1985) and in Machenhauer (1991).

One recent idea for removing pole problems for FD methods involves mapping each hemisphere (plus some additional equatorial region) separately to (singularity-free) 2-D flat domains. Standard FD methods can be used on each of the two domains, which are coupled through interpolation in the areas of overlap. Browning, Hack, and Swarztrauber (1988) find this approach (using FD4 or FD6 approximations) to be roughly four times *slower* than spherical harmonics-based PS methods for some model problems relevant to meteorology.

Global climate codes are related to global weather codes, but with some significant differences that might favor spectral techniques:

- much longer time scales; and
- less irregular fine-scale forcing terms (e.g., no need to model in detail the quite strong irregularities associated with each local instance of precipitation).

Limited area models (LAMs) are commonly implemented as local high-resolution corrections applied to global forecasts. Because long-distance interactions grow in importance over time (for example, the interactions between the hemispheres become important after about 5–6 days), LAMs usually extend over shorter times than global forecasts. As in most applications, the presence of domain boundaries causes particular problems for spectral methods. However, Fourier-spectral LAMs became operational at the Japan Meteorological Agency in 1986 and ECMWF in 1987. These codes artificially imposed periodicity over rectangular domains. Haugen and Machenhauer (1993) describe a Fourier spectral method that uses "correction layers" surrounding the domain of interest (and give references to the other spectral LAMs just mentioned).

In conclusion, we should note that codes for weather forecasting are very complex. Spectral codes contain of course significant spectral components, but also many other numerical techniques better suited to representing some of the meteorological effects of interest.

8.4. Seismic exploration

In contrast to many other mineral deposits, hydrocarbon (e.g. natural gas or oil) reservoirs do not produce any distinctive electric, magnetic, or gravitational anomalies that can be detected from the surface. Apart from drilling (which is very expensive), the only well established and widely applicable exploration procedure is to use seismic waves. Such waves are:

- easy to generate (explosions, truck-mounted vibrators, air guns towed after ships, etc.);
- able to penetrate many kilometers of earth, water, and rock with only moderate energy loss;
- strongly influenced by properties of the medium – in particular, interfaces cause reflections that bring relatively strong return signals back to the surface; and
- easy to detect on their return to the surface.

In most places where gas and oil were once present, these have long ago risen to the surface (and biodegraded). Although seismic imaging is unable to show hydrocarbons directly, it can show geological formations that might have been able to trap such hydrocarbons on their way up.

Ideas from geometrical optics can be used to follow seismic-ray paths if the wavelengths are short compared to the size of the layers that are studied (a questionable assumption – both are often in the range of 10–200 meters). In the case of interfaces, energy is transferred between different types of waves. Part of the energy is transmitted and part is reflected. Even with only a few layers, a vast number of waves will arise. The goal is to somehow invert the surface response data into a picture of the underground structures. The crudest possible method of finding the depth of the first interface would be to measure only the travel time for the first reflected signal. To obtain a more complete picture, a large number of corrections and refinements must be made, including:

normal move-out – correction for the varying thickness of the weathered layer nearest the surface;
stacking – combining data from many source and receiver positions to reduce noise and cancel effects from multiple reflections; and
migration – a correction procedure for inclined interfaces.

The resulting seismic images provide skilled exploration geophysicists with a rough picture of the major interfaces. Using this as a first approximation, the elastic wave equation can then be solved with FD (or FE or PS) methods to generate synthetic seismograms. From the differences between these and real seismograms, one can iteratively change the assumed structures to bring the synthetic responses closer to the real ones.

Elastic wave equation

For the relatively low-amplitude waves that are used, the earth can be viewed as a linear, elastic medium. Such media can support several types of waves.

P waves – "P" stands for pressure or primary. These are longitudinal (sound) waves.

S waves – "S" denotes shear or secondary; these are transversal waves.

Rayleigh waves – waves following a surface, which decay in strength exponentially down into the ground. (Rayleigh waves are often the main cause of earthquake damage, since they keep their energy along the surface.)

Stoneley waves – waves following an interface between different media. These decay exponentially in both directions away from the interface.

Love waves – waves trapped between interfaces, traveling as in a wave guide.

The equations of motion of an elastic material under small deformations were established in the first half of the 19th century by Poisson, Navier, Cauchy, Stokes, Green, and others. A particularly simple formulation of these equations in the case of a 2-D isotropic medium was given in Figure 5.3-1:

$$\begin{aligned} \rho u_t &= f_x + g_y, \\ \rho v_t &= g_x + h_y, \\ f_t &= (\lambda+2\mu)u_x + \lambda v_y, \\ g_t &= \mu v_x + \mu u_y, \\ h_t &= \lambda u_x + (\lambda+2\mu)v_y, \end{aligned} \tag{8.4-1}$$

where: u, v are velocities in the x and the y direction (resp.); f, g, h are stress components; and ρ, λ, μ are material parameters – given functions of x and y (for fluids, $\mu = 0$).

This first-order system correctly propagates all the types of elastic waves (e.g., P and S waves with speeds $\sqrt{(\lambda+2\mu)/\rho}$ and $\sqrt{\mu/\rho}$ resp.), gives correct reflections and transmissions at interfaces, includes all effects of diffraction at edges, and so forth. Wave attenuation is not included, but can easily be added. The 2-D forward modeling problem consists of solving equations such as (8.4-1) in rectangular domains with a free surface on top and absorbing (nonreflecting) boundary conditions on the other three sides.

Numerical solution of the elastic wave equation

Finite difference modeling of wave equations in seismic contexts began in earnest around 1970. Alford, Kelly, and Boore (1974) note that FD4 is superior to FD2 for the 2-D acoustic ($\mu = 0$) wave equation. Another early paper of lasting interest is Kelly et al. (1976), which shows how FD modeling can aid seismic interpretation in cases of complex geometries. Many production codes for solving the elastic wave equation were upgraded from FD2 to FD4 around 1986.

Although PS methods have not yet become widely used in production codes, they are receiving much attention in the seismic exploration literature. Kosloff and Baysal (1982) test a Fourier–PS method on two 2-D acoustic model problems and find it more effective than an FD scheme. Johnson (1984) applies the Fourier–PS method to a 3-D acoustic problem, and Kosloff, Reshef, and Loewenthal (1984) to a 2-D elastic problem. After comparing FD2 and FD4 against Fourier–PS on a number of 2-D elastic problems (Fornberg 1987; one test case was shown in Figure 7.1-2), it became clear that PS methods would decidedly outperform FD2 and FD4 methods if two main difficulties – present for FD methods, but more severe for PS methods – could be resolved.

(1) Treatment of *outflow boundary conditions:* Many waves that leave the source in directions away from the receiver will not influence the return signals during the time of interest. For efficiency, the computational domain must be kept small in size. The problem is to implement artificial boundaries so that they do not reflect back any unwanted signals.

(2) Representation of *interfaces:* To specify material properties at all grid-point locations leaves grid-size uncertainties in interface locations – and causes gently sloping interfaces to appear jagged.

Outflow boundary conditions: The "classical" method against which later progress is measured is quite recent: Engquist and Majda (1977) and Clayton and Engquist (1977) describe how *pseudodifferential operators* can be devised that allow wave propagation only in certain directions. FD versions of these can be applied as boundary conditions, thus allowing waves to pass out of the domain but not be reflected back in. Many variations of this approach have been described; see for example Halpern and Trefethen (1988), Long and Liow (1990), Higdon (1990, 1991), and Tirkas, Balanis, and Renaut (1992).

In some instances, boundary conditions that are effective on absorbing (physically relevant) outgoing waves have been found to allow instabilities of the type shown in Figure 4.5-3(b).

One-way wave approximations can work well for FD and Chebyshev–PS methods, but not for Fourier–PS methods (which do not accept boundaries at any locations). Cerjan et al. (1985) and Kosloff and Kosloff (1986) discuss the technique of enlarging the computational domain and using the additional portion to

- smoothly join up the physical domain sides to create a periodic problem, and
- apply a damping mechanism sufficiently strong that very little energy passes through, but still sufficiently weak that equally little energy is reflected back.

Renaut and Frölich (1995) combine this with the grid stretching proposed by Kosloff and Tal-Ezer (1993) (and described in Section 5.5).

Interfaces: A gently sloping or curved interface will, on a grid, look identical to an interface with piecewise horizontal and vertical segments halfway between grid lines. Experiments by Fornberg (1988a) show that FD and, in particular, Fourier–PS methods will handle the interface quite accurately under the latter assumption. This has both good and bad consequences:

+ If only a few weakly curved and largely parallel interfaces are present, it may be possible to perform a coordinate transformation that aligns the grid with the interfaces. Very high computational accuracy can then be achieved at low cost.
– If a wavefront passes over perceived "kinks", these will act as spurious point sources for radially emanating elastic waves.

Figure 8.4-1(a) shows a test problem with a straight P wavefront traveling down into a curved interface. Figure 8.4-1(b) shows an accurate numerical solution after the wave has passed down through the interface and generated a weak reflected wave traveling back up. Figure 8.4-1(c)

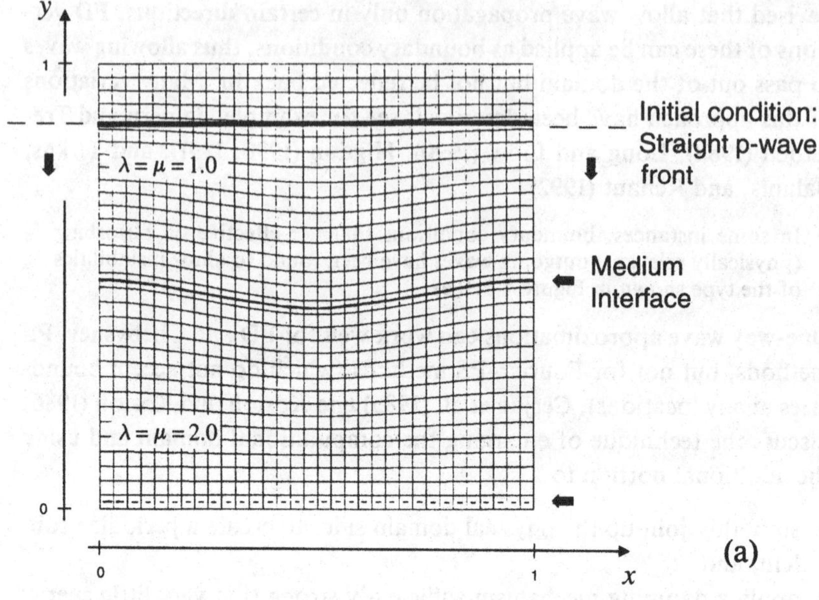

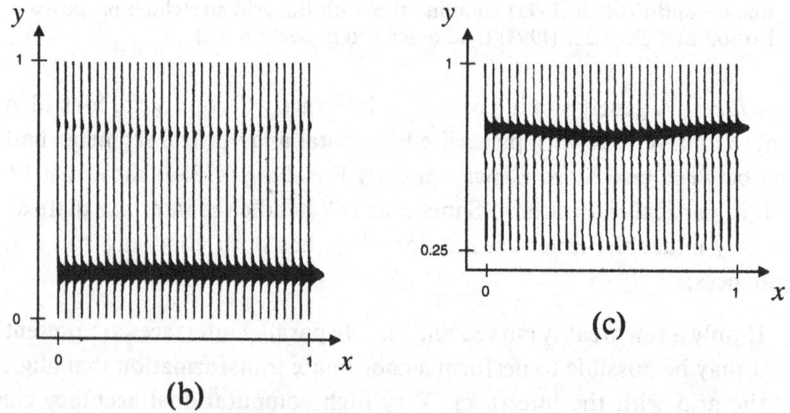

Figure 8.4-1. **(a)** Initial condition, structure of medium and mapped grid. **(b)** Numerical solution (variable f) some time after wavefront has partly passed through and partly reflected back from the interface. **(c)** Top three-quarters of part (b). The amplitude scale is adjusted to display more detail.

shows the top three-fourths of Figure 8.4-1(b), but with the display scale for the amplitude changed (to allow more details to be seen). Figure 8.4-2 shows the equivalent of Figure 8.4-1(c) when different methods and grid densities are used. The artificial point sources are overwhelmed by other

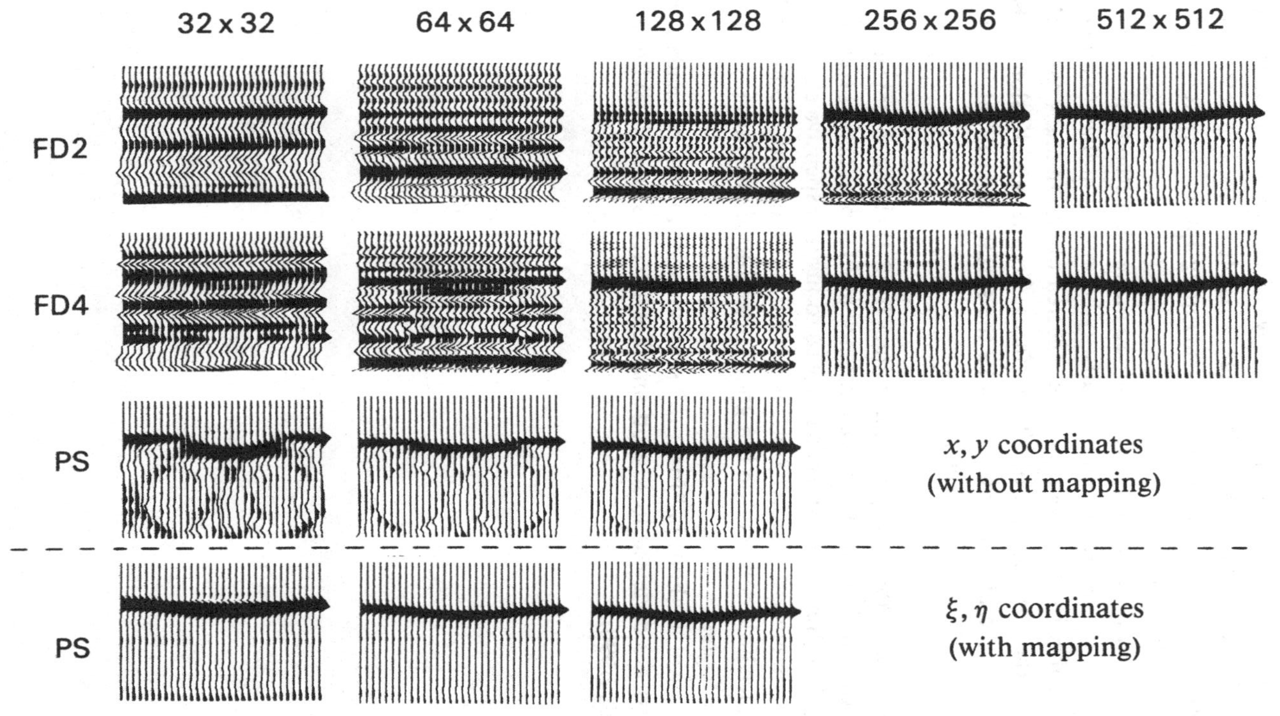

Figure 8.4-2. Same situation as in Figure 8.4-1(c), here computed with different methods and grid densities.

errors in the case of the FD2 method. They can barely be seen when using FD4 (on the 128×128 grid), but are highly visible for the PS method (because other errors are very small). The mapping approach restores excellent PS performance.

No good method has yet been found to make PS methods interpret an interface as located anywhere but halfway between grid lines (without a major sacrifice in the accuracy of the reflected wave). In many cases, the surface is the most critical of all interfaces. Tessmer and Kosloff (1994) propose the use of a mapping to make the surface appear flat in a 3-D elastic Chebyshev–PS implementation.

Present Fourier and Chebyshev techniques for 2-D elastic seismic modeling perform with about the same efficiency as FD4 production codes (or marginally better – but not sufficiently so to justify expensive changes in codes). The present industry trend is toward 3-D modeling. PS methods would face little competition if a good subgrid-level interface procedure could be found.

Numerical methods in "forward" seismic modeling (in particular, PS methods) are reviewed in Kosloff and Kessler (1989) and in Witte and Richards (1990).

Appendices

A
Jacobi polynomials

The Jacobi polynomials $P_n^{\alpha,\beta}(x)$ of degree n, $n=0,1,2,\ldots$, depend on parameters α and β (both >-1). The most immediate way to define these polynomials is through their orthogonality relation:

$$\int_{-1}^{1} P_n^{\alpha,\beta}(x)P_m^{\alpha,\beta}(x)(1-x)^\alpha(1+x)^\beta\,dx=0 \quad \text{for } m\neq n.$$

This relation is critical to their role in Gaussian quadrature formulas, but is quite unrelated to why the polynomials also happen to be useful in connection with PS methods. This latter topic is treated throughout Chapters 2–4. The purpose of this appendix is to support that discussion by providing a convenient collection of key formulas. General theoretical background as well as further relations (and proofs) can be found in numerous books on orthogonal polynomials, such as Sansone (1959), Szegö (1959), and Luke (1969).

The two most important special cases of Jacobi polynomials are Legendre and Chebyshev polynomials, obtained (resp.) by setting $\alpha=\beta=0$ and $\alpha=\beta=-\frac{1}{2}$. The Chebyshev case is often preferred for numerical work because:

(1) numerous particularly simple closed-form expressions and relations are available to simplify the algebra;
(2) conversions between the coefficients a_k in a Chebyshev expansion $\sum_{k=0}^{n} a_k T_k(x)$ and function values at the Chebyshev nodes (extrema of $T_n(x)$) can be performed especially rapidly for large n by means of the fast cosine transform (FCT - see Appendix F); and
(3) use of the Chebyshev nodes makes polynomial interpolation marginally more robust (see the discussions on Lebesgue constants in Section 2.1 and Appendix D).

Table A-1. *Jacobi polynomial fact sheet*

	Legendre	Chebyshev	Jacobi
Weight function $W(x)$	1	$\frac{1}{\sqrt{1-x^2}}$	$(1-x)^\alpha(1+x)^\beta$ $\alpha>-1,\ \beta>-1$ $\alpha=\beta=0$ (Legendre) $\alpha=\beta=-\frac{1}{2}$ (Chebyshev)
Customary normalization	$P_n(1)=1$	$T_n(1)=1$	$P_n^{\alpha,\beta}(1)=\binom{n+\alpha}{n}$
First few polynomials	1 x $\frac{3}{2}x^2-\frac{1}{2}$ $\frac{5}{2}x^3-\frac{3}{2}x$ $\frac{35}{8}x^4-\frac{15}{4}x^2+\frac{3}{8}$ $\frac{63}{8}x^5-\frac{35}{4}x^3+\frac{15}{8}x$	1 x $2x^2-1$ $4x^3-3x$ $8x^4-8x^2+1$ $16x^5-20x^3+5x$	1 $\frac{1}{2}(2+\alpha+\beta)x+\frac{1}{2}(\alpha-\beta)$ $\frac{1}{8}(3+\alpha+\beta)(4+\alpha+\beta)x^2+\frac{1}{4}(\alpha-\beta)(3+\alpha+\beta)x+\frac{1}{8}[(\alpha-\beta)^2-(4+\alpha+\beta)]$ General n: $2^{-n}\sum_{k=0}^{n}\binom{n+\alpha}{k}\binom{n+\beta}{n-k}(x-1)^{n-k}(x+1)^k$

Orthogonality $\int_{-1}^{1} \phi_m \phi_n W\,dx$	$0 \quad (m \neq n)$ $\frac{2}{2n+1} \quad (m=n)$	$0 \ (m \neq n)$ $\pi \ (m=n=0)$ $\frac{\pi}{2} \ (m=n>0)$	$0 \ (m \neq n)$ $\frac{2^{\alpha+\beta+1}\Gamma(n+\alpha+1)\Gamma(n+\beta+1)}{(2n+\alpha+\beta+1)n!\,\Gamma(n+\alpha+\beta+1)} \ (m=n)$
Three-term recursion	$(n+1)P_{n+1}$ $-(2n+1)xP_n$ $+nP_{n-1}=0$	T_{n+1} $-2xT_n$ $+T_{n-1}=0$	$2(n+1)(n+\alpha+\beta+1)(2n+\alpha+\beta)P_{n+1}$ $-[(2n+\alpha+\beta+1)(\alpha^2-\beta^2)$ $+(2n+\alpha+\beta)(2n+\alpha+\beta+1)(2n+\alpha+\beta+2)x]P_n$ $+2(n+\alpha)(n+\beta)(2n+\alpha+\beta+2)P_{n-1}=0$
Differential equation	$(1-x^2)P_n''-2xP_n'$ $+n(n+1)P_n=0$	$(1-x^2)T_n''-xT_n'$ $+n^2T_n=0$	$(1-x^2)P_n''$ $+[(\beta-\alpha)-(\alpha+\beta+2)x]P_n'+n(n+\alpha+\beta+1)P_n=0$
First derivative recursion	$P_{n+1}'-P_{n-1}'$ $=(2n+1)P_n$	$\frac{T_{n+1}'}{n+1}-\frac{T_{n-1}'}{n-1}$ $=2T_n$	$2(2n+\alpha+\beta)(n+\alpha+\beta)(n+\alpha+\beta+1)P_{n+1}'$ $+2(\alpha-\beta)(n+\alpha+\beta)(2n+\alpha+\beta+1)P_n'$ $-2(n+\alpha)(n+\beta)(2n+\alpha+\beta+2)P_{n-1}'$ $=(n+\alpha+\beta)(2n+\alpha+\beta)(2n+\alpha+\beta+1)(2n+\alpha+\beta+2)P_n$

Table A-2. *Chebyshev polynomial fact sheet*

Explicit expressions	$T_n(x) = \cos n\theta, \quad \theta = \arccos x$	←
	$= \frac{1}{2}\left(z^n + \frac{1}{z^n}\right)$ where $\frac{1}{2}\left(z + \frac{1}{z}\right) = x$	←
	$= \frac{1}{2}[(x+\sqrt{x^2-1})^n + (x-\sqrt{x^2-1})^n]$	←
	$= \frac{n}{2}\sum_{k=0}^{[n/2]}(-1)^k \frac{(n-k-1)!}{k!\,(n-2k)!}(2x)^{n-2k}$ where $[n/2]$ = integer part of $n/2$ (for $n \neq 0$; $T_0(x) \equiv 1$)	
	$= \frac{(-1)^k\sqrt{\pi}}{2^n\Gamma(n+\frac{1}{2})}\sqrt{1-x^2}\,\frac{d^n}{dx^n}[(1-x^2)^{n-1/2}]$ (Rodrigues's formula)	
Zeros	$T_n(x_k) = 0$ for $x_k = \cos\frac{2k-1}{2n}\pi, \ k = 1, 2, \ldots, n$	←
Extrema	$x_k = \cos\frac{k\pi}{n}, \ k = 0, 1, \ldots, n$	←
Miscellaneous formulas	$2T_m(x)T_n(x) = T_{n+m}(x) + T_{n-m}(x) \ (n \geq m)$	←
	$\frac{1-xz}{1-2xz+z^2} = \sum_{k=0}^{\infty} T_k(x)z^n$ (generating function)	
	$(1-x^2)T_n'(x) = -nxT_n(x) + nT_{n-1}(x)$	

Notes: This table lists relations that are additional to those shown in Table A-1. Arrows at the right indicate results that lack simple counterparts for other Jacobi polynomials.

Table A-1 summarizes some general relations for Jacobi polynomials and lists their simplified forms in the Legendre and Chebyshev cases.

Many relations for Jacobi polynomials can be obtained from the large number of identities that are known for hypergeometric functions. The two classes of functions are related through

$$P_n^{(\alpha,\beta)}(1-2z) = \frac{(\alpha+1)_n}{n!}F(-n, n+\alpha+\beta+1; \alpha+1; z),$$

where $(\alpha+1)_n = (\alpha+1)\cdot(\alpha+2)\cdots(\alpha+n)$.

Table A-2 lists a number of additional relations satisfied by Chebyshev polynomials. Arrows in the right margin indicate results that lack simple generalizations to the other cases. Figure A-1 compares some Legendre and Chebyshev polynomials over $[-1, 1]$.

All Jacobi polynomials of degree n have their n zeros in the *fundamental* interval $[-1, 1]$, clustered quadratically toward ± 1; this is a consequence of

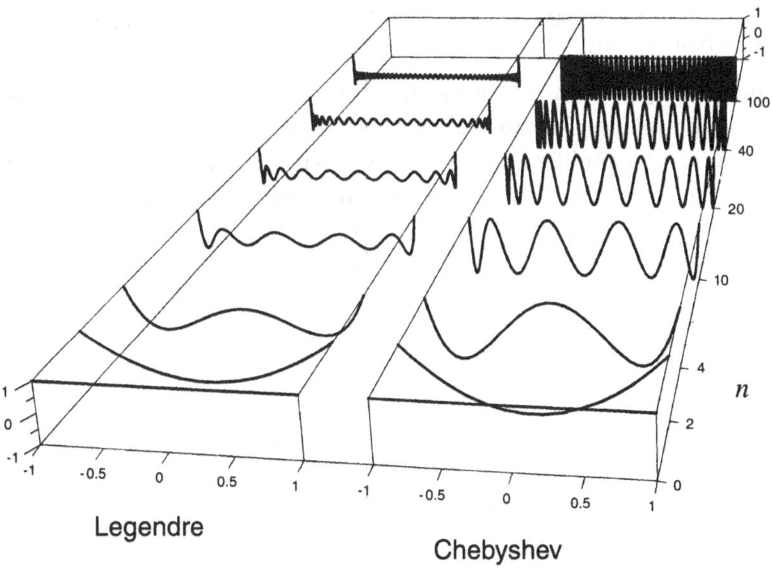

Figure A-1. Comparison of Legendre and Chebyshev polynomials on $[-1, 1]$.

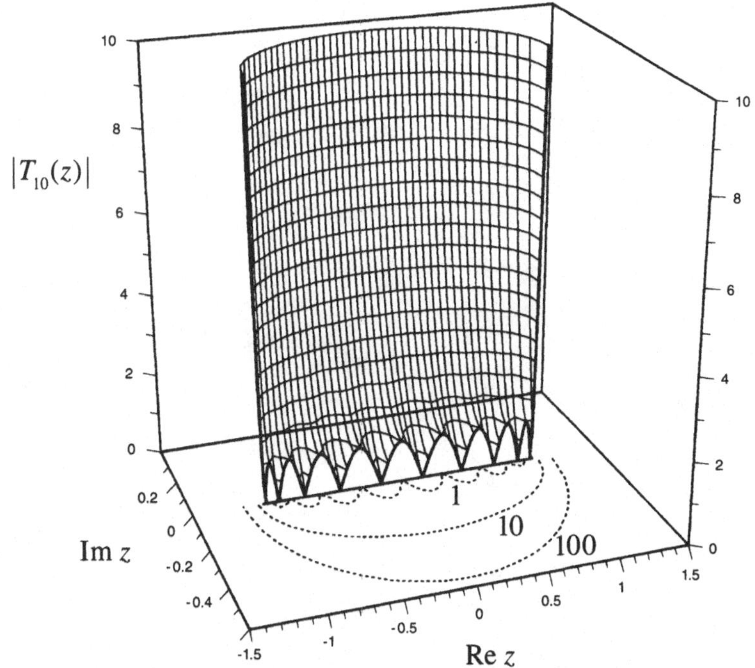

Figure A-2. Growth of $|T_{10}(z)|$ in the complex plane. The result is symmetric in both axes – the front half of the figure shows only some contour lines.

their differential equation (Table A-1). The significance of this is discussed in Sections 3.3 and 3.4.

In the complex plane outside $[-1, 1]$, Jacobi polynomials diverge to infinity very rapidly. Contour curves $|P_n^{\alpha,\beta}(z)| = M$ for either M or n (or both) increasing rapidly approach ellipses with foci at ± 1. Figure A-2 illustrates this growth in the case of the Chebyshev polynomial $T_{10}(z)$.

B

Tau, Galerkin, and collocation (PS) implementations

We consider the model problem

$$u_{xx}+u_x-2u+2=0, \quad -1\le x\le 1,$$
$$u(-1)=u(1)=0,$$

and approximate the exact solution

$$u(x)=1-\frac{\sinh(2)e^{x}+\sinh(1)e^{-2x}}{\sinh(3)}$$

by a truncated Chebyshev expansion

$$v(x)=\sum_{k=0}^{4} a_k T_k(x).$$

From (2.1-1) and (2.1-3) it follows that the residual

$$R(x)=v_{xx}+v_x-2v+2=\sum_{k=0}^{4} A_k T_k(x) \tag{B-1}$$

satisfies

$$\begin{bmatrix} A_0\\ A_1\\ A_2\\ A_3\\ A_4 \end{bmatrix}=\begin{bmatrix} -2 & 1 & 4 & 3 & 32\\ 0 & -2 & 4 & 24 & 8\\ 0 & 0 & -2 & 6 & 48\\ 0 & 0 & 0 & -2 & 8\\ 0 & 0 & 0 & 0 & -2 \end{bmatrix}\begin{bmatrix} a_0\\ a_1\\ a_2\\ a_3\\ a_4 \end{bmatrix}+\begin{bmatrix} 2\\ 0\\ 0\\ 0\\ 0 \end{bmatrix}. \tag{B-2}$$

The matrix in (B-2) is obtained as

$$\begin{bmatrix} 0 & 0 & 4 & 0 & 32\\ & 0 & 0 & 24 & 0\\ & & 0 & 0 & 48\\ & & & 0 & 0\\ & & & & 0 \end{bmatrix}+\begin{bmatrix} 0 & 1 & 0 & 3 & 0\\ & 0 & 4 & 0 & 8\\ & & 0 & 6 & 0\\ & & & 0 & 8\\ & & & & 0 \end{bmatrix}-2\begin{bmatrix} 1 & & & & \\ & 1 & & & \\ & & 1 & & \\ & & & 1 & \\ & & & & 1 \end{bmatrix},$$

corresponding (resp.) to v_{xx}, v_x, and $-2v$. The matrix A for v_x can be obtained as

$$v_x \Leftrightarrow A = \begin{bmatrix} 0 & 1 & 0 & 3 & 0 \\ & 0 & 4 & 0 & 8 \\ & & 0 & 6 & 0 \\ & & & 0 & 8 \\ & & & & 0 \end{bmatrix} = \begin{bmatrix} 0 & \begin{bmatrix} 1 & 0 & -\frac{1}{2} & \\ & \frac{1}{4} & 0 & -\frac{1}{4} \\ & & \frac{1}{6} & 0 \\ & & & \frac{1}{8} \end{bmatrix}^{-1} \\ 0 & 0 \quad 0 \quad 0 \quad 0 \end{bmatrix} = \begin{bmatrix} 0 & [C]^{-1} \\ 0 & 0 \end{bmatrix},$$

where C is the same matrix as in (2.1-3). There are (at least) three ways to compute the matrices A^p corresponding to higher derivatives d^pv/dx^p.

1. Form the matrix powers A^p directly from A by repeated matrix multiplications.
2. Instead of repeated multiplications by A, repeatedly back-substitute (column by column) using the tri-diagonal matrix C.
3. Use explicit formulas for A and A^2:

$$A_{i,j} = \begin{cases} (1/c_i) \times 2j & \text{if } j > i, \ i+j \text{ odd}, \\ 0 & \text{otherwise}, \end{cases}$$

$$A^2_{i,j} = \begin{cases} (1/c_i) \times (j-i)j(j+i) & \text{if } j > i, \ i+j \text{ even}, \\ 0 & \text{otherwise}, \end{cases}$$

where

$$0 \le i,j \le n \quad \text{and} \quad c_i = \begin{cases} 2 & \text{if } i = 0, \\ 1 & \text{if } i > 0, \end{cases}$$

followed by the recursion

$$A^{p+2}_{i,j} = A^p_{i,j} \cdot \frac{(j-i-p)(j+i+p)(j-i+p)(j+i-p)}{4p(p+1)}, \quad p \ge 1.$$

The last two methods are particularly fast numerically. All three methods generalize from Chebyshev to general Jacobi polynomials. For example, in the case of Legendre polynomials, method 3 becomes

$$A_{i,j} = \begin{cases} 2i+1 & \text{if } j > i, \ i+j \text{ odd}, \\ 0 & \text{otherwise}, \end{cases}$$

$$A^2_{i,j} = \begin{cases} (i+\frac{1}{2})(j-i)(j+i+1) & \text{if } j > 1, \ i+j \text{ even}, \\ 0 & \text{otherwise}, \end{cases}$$

and

$$A^{p+2}_{i,j} = A^p_{i,j} \cdot \frac{(j-i-p)(j+i+1+p)(j-i+p)(j+i+1-p)}{4p(p+1)}, \quad p \ge 1.$$

The two recursions shown for method 3 can be proven in a similar manner as the (nonrecursive) formulas in Karageorghis (1988) and Phillips (1988), respectively.

Enforcing the boundary conditions $v(-1) = v(1) = 0$ leads to

$$\begin{bmatrix} 1 & 1 & 1 & 1 & 1 \\ 1 & -1 & 1 & -1 & 1 \end{bmatrix} \begin{bmatrix} a_0 \\ a_1 \\ a_2 \\ a_3 \\ a_4 \end{bmatrix} = \begin{bmatrix} 0 \\ 0 \end{bmatrix}. \tag{B-3}$$

Ideally, we would like to make $A_i = 0$, $i = 0, 1, \ldots, 4$, while still satisfying (B-3). However, this would mean satisfying seven relations with only five free parameters a_i, $i = 0, 1, \ldots, 4$. The three spectral methods differ in how they approximate this overdetermined system.

Tau. Require $R(x)$ in (B-1) to be orthogonal to $T_k(x)$, $k = 0, 1, 2$:

$$\int_{-1}^{1} \frac{R(x)T_k(x)}{\sqrt{1-x^2}}\,dx = 0 \Rightarrow \begin{cases} A_0 = 0, \\ A_1 = 0, \\ A_2 = 0. \end{cases}$$

The top three lines of (B-2) together with (B-3) yield

$$[a_0, \ldots, a_4] = [.2724, -.0444, -.2562, .0444, -.0162].$$

Galerkin. Create from $T_0, \ldots, T_4$ three basis functions ϕ_2, ϕ_3, ϕ_4 satisfying both boundary conditions,

$$\begin{aligned} \phi_2(x) &= T_2(x) - T_0(x), \\ \phi_3(x) &= T_3(x) - T_1(x), \\ \phi_4(x) &= T_4(x) - T_0(x), \end{aligned}$$

and require $R(x)$ to be orthogonal to $\phi_k(x)$, $k = 2, 3, 4$:

$$\int_{-1}^{1} \frac{R(x)\phi_k(x)}{\sqrt{1-x^2}}\,dx = 0 \Rightarrow \begin{bmatrix} 2 & 0 & -1 & 0 & 0 \\ 0 & 1 & 0 & -1 & 0 \\ 2 & 0 & 0 & 0 & -1 \end{bmatrix} \begin{bmatrix} A_0 \\ A_1 \\ A_2 \\ A_3 \\ A_4 \end{bmatrix} = \begin{bmatrix} 0 \\ 0 \\ 0 \end{bmatrix}.$$

Together with (B-2) and (B-3), this yields

$$[a_0, \ldots, a_4] = [.2741, -.0370, -.2593, .0370, -.0148]$$

Collocation (PS). Force $R(x_i) = 0$ at $x_i = \cos(i\pi/4)$, $i = 1, 2, 3$:

$$\begin{bmatrix} 1 & \frac{1}{\sqrt{2}} & 0 & -\frac{1}{\sqrt{2}} & -1 \\ 1 & 0 & -1 & 0 & 1 \\ 1 & -\frac{1}{\sqrt{2}} & 0 & \frac{1}{\sqrt{2}} & -1 \end{bmatrix} \begin{bmatrix} A_0 \\ A_1 \\ A_2 \\ A_3 \\ A_4 \end{bmatrix} = \begin{bmatrix} 0 \\ 0 \\ 0 \end{bmatrix}.$$

The section of the discrete cosine transform matrix has entries $T_k(x_i) = \cos(ki\pi/4)$, $k = 0, 1, \ldots, 4$, $i = 1, 2, 3$. Combining this with (B-2) and (B-3), we have

$$[a_0, \ldots, a_4] = [.2743, -.0371, -.2600, .0371, -.0143].$$

In exact arithmetic, $[a_0, \ldots, a_4] = [\frac{48}{175}, -\frac{13}{350}, -\frac{13}{50}, \frac{13}{350}, -\frac{1}{70}]$ and the values at the node locations x_i, $i = 0, \ldots, 4$, become

$$[0, \tfrac{101}{350} + \tfrac{13}{350}\sqrt{2}, \tfrac{13}{25}, \tfrac{101}{350} - \tfrac{13}{350}\sqrt{2}, 0].$$

This description of the PS approach followed the style used to describe the tau and Galerkin methods, but gave no indication why the PS approach is more flexible than the other two in cases of variable coefficients and nonlinearities. We therefore describe the PS method again, this time in terms of nodal values rather than expansion coefficients.

Collocation (PS) – FD-based description. If v_i denotes the approximations at the nodes $x_i = \cos(i\pi/4)$, $i = 0, 1, \ldots, 4$, the first and second derivative of the interpolating polynomial take (at the node locations) the values

$$\begin{bmatrix} v_{x\,0} \\ v_{x\,1} \\ v_{x\,2} \\ v_{x\,3} \\ v_{x\,4} \end{bmatrix} = \begin{bmatrix} -\frac{11}{2} & 4+2\sqrt{2} & -2 & 4-2\sqrt{2} & -\frac{1}{2} \\ -1-\frac{1}{2}\sqrt{2} & \frac{1}{2}\sqrt{2} & \sqrt{2} & -\frac{1}{2}\sqrt{2} & 1-\frac{1}{2}\sqrt{2} \\ \frac{1}{2} & -\sqrt{2} & 0 & \sqrt{2} & -\frac{1}{2} \\ -1+\frac{1}{2}\sqrt{2} & \frac{1}{2}\sqrt{2} & -\sqrt{2} & -\frac{1}{2}\sqrt{2} & 1+\frac{1}{2}\sqrt{2} \\ \frac{1}{2} & -4+2\sqrt{2} & 2 & -4-2\sqrt{2} & \frac{11}{2} \end{bmatrix} \begin{bmatrix} v_0 \\ v_1 \\ v_2 \\ v_3 \\ v_4 \end{bmatrix} \tag{B-4}$$

and

$$\begin{bmatrix} v_{xx\,0} \\ v_{xx\,1} \\ v_{xx\,2} \\ v_{xx\,3} \\ v_{xx\,4} \end{bmatrix} = \begin{bmatrix} 17 & -20-6\sqrt{2} & 18 & -20+6\sqrt{2} & 5 \\ 5+3\sqrt{2} & -14 & 6 & -2 & 5-3\sqrt{2} \\ -1 & 4 & -6 & 4 & -1 \\ 5-3\sqrt{2} & -2 & 6 & -14 & 5+3\sqrt{2} \\ 5 & -20+6\sqrt{2} & 18 & -20-6\sqrt{2} & 17 \end{bmatrix} \begin{bmatrix} v_0 \\ v_1 \\ v_2 \\ v_3 \\ v_4 \end{bmatrix} \tag{B-5}$$

respectively.

These are examples of *differentiation matrices,* which are discussed in numerous places in this review. Each matrix row contains the weights of an FD stencil that is as wide as the grid is wide. Section 4.3 describes how the matrix elements can be obtained very conveniently.

Enforcing

$$R(x) = v_{xx} + v_x - 2v + 2 = 0$$

at the node points x_k, $k = 1, 2 = 3$, and enforcing the boundary conditions $v_0 = v_4 = 0$ lead to

$$\begin{bmatrix} -16+\frac{1}{2}\sqrt{2} & 6+\sqrt{2} & -2-\frac{1}{2}\sqrt{2} \\ 4-\sqrt{2} & -8 & 4+\sqrt{2} \\ -2+\frac{1}{2}\sqrt{2} & 6-\sqrt{2} & -16-\frac{1}{2}\sqrt{2} \end{bmatrix} \begin{bmatrix} v_1 \\ v_2 \\ v_3 \end{bmatrix} = \begin{bmatrix} -2 \\ -2 \\ -2 \end{bmatrix}, \tag{B-6}$$

with the same solution as before:

$$\begin{bmatrix} v_1 \\ v_2 \\ v_3 \end{bmatrix} = \begin{bmatrix} \frac{101}{350} + \frac{13}{350}\sqrt{2} \\ \frac{13}{25} \\ \frac{101}{350} - \frac{13}{350}\sqrt{2} \end{bmatrix}.$$

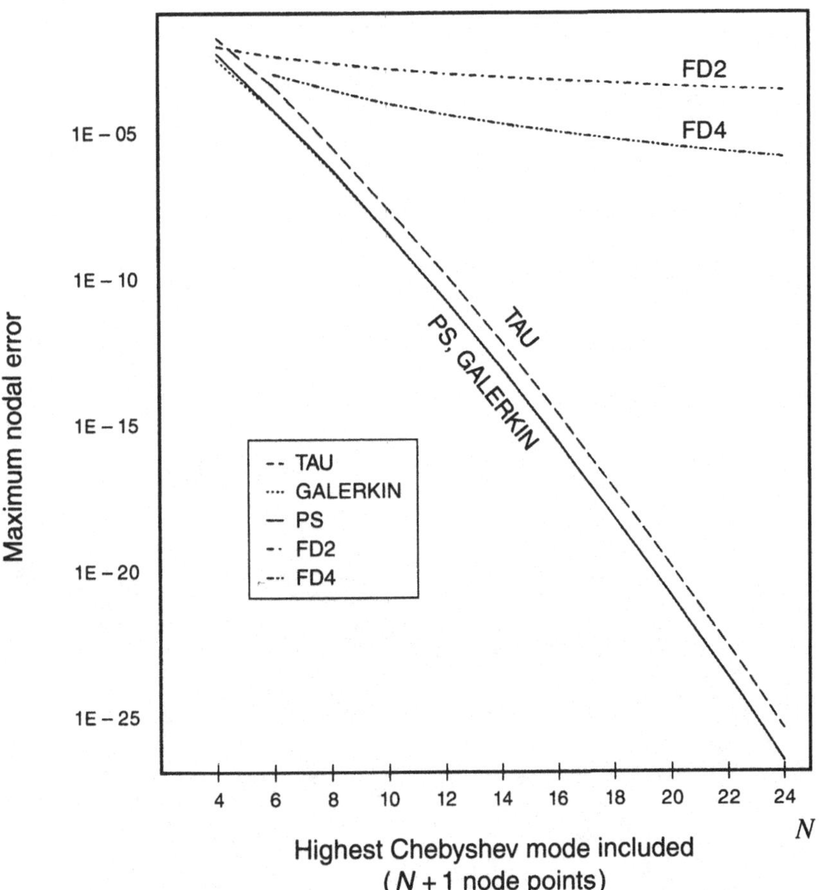

Figure B-1. Maximum nodal errors for different methods when applied to the model in Appendix B: comparison among three spectral implementations and equi-spaced FD methods of second and fourth order.

Had variable coefficients been present in the governing equation, the only modification would have been to multiply – when assembling (B-6) – the rows of (B-4) and (B-5) with the values at the nodes.

Figure B-1 shows how the accuracy increases with n: all three cases feature an exponential rate of convergence. For comparison, curves for FD2 and FD4 (on equi-spaced grids) are also included.

C

Codes for algorithm to find FD weights

This appendix gives two codes (and a test driver) for the algorithm that was described in Section 3.1. The first code (`WEIGHTS`) returns weights for all FD stencils of widths up to (and including) the maximum specified. The second one (`WEIGHTS1`) returns only the widest (most accurate) approximation for each derivative.

```
      SUBROUTINE WEIGHTS (XI,X,N,M,C)
C +----------------------------------------------------------------+
C |  INPUT PARAMETERS:                                             |
C |    XI   POINT AT WHICH THE APPROXIMATIONS ARE TO BE ACCURATE   |
C |    X    X-COORDINATES FOR GRID POINTS, ARRAY DIMENSIONED X(0:N)|
C |    N    THE GRID POINTS ARE AT X(0),X(1),...,X(N) (I.E. N+1 IN ALL)|
C |    M    HIGHEST ORDER OF DERIVATIVE TO BE APPROXIMATED         |
C |                                                                |
C |  OUTPUT PARAMETER:                                             |
C |    C    WEIGHTS, ARRAY DIMENSIONED C(0:N,0:N,0:M).             |
C |         ON RETURN, THE ELEMENT C(J,I,K) CONTAINS THE WEIGHT TO BE |
C |         APPLIED AT X(J) WHEN THE K:TH DERIVATIVE IS APPROXIMATED |
C |         BY A STENCIL EXTENDING OVER X(0),X(1),...,X(I).        |
C +----------------------------------------------------------------+
      IMPLICIT REAL*8 (A-H,O-Z)
      DIMENSION X(0:N),C(0:N,0:N,0:M)
      C(0,0,0) = 1.0D0
      C1       = 1.0D0
      C4       = X(0)-XI
      DO 40 I=1,N
         MN    = MIN(I,M)
         C2    = 1.0D0
         C5    = C4
         C4    = X(I)-XI
         DO 20 J=0,I-1
            C3 = X(I)-X(J)
            C2 = C2*C3
            IF (I.LE.M) C(J,I-1,I)=0.0D0
            C(J,I,0) = C4*C(J,I-1,0)/C3
            DO 10 K=1,MN
```

```
10             C(J,I,K) = (C4*C(J,I-1,K)-K*C(J,I-1,K-1))/C3
20          CONTINUE
         C(I,I,0) = -C1*C5*C(I-1,I-1,0)/C2
         DO 30 K=1,MN
30          C(I,I,K) = C1*(K*C(I-1,I-1,K-1)-C5*C(I-1,I-1,K))/C2
40       C1 = C2
      RETURN
      END
```

Note: If N is very large, the calculation of the variable C2 might cause overflow (or underflow). For example, in generating extensions of Tables 3.1-1 and 3.1-2, this problem arises when $N!$ exceeds the largest possible number – that is, $N > 34$ in typical 32-bit precision with $3 \cdot 10^{38}$ as the largest number; $N > 965$ in CRAY single precision (64-bit word length, 15-bit exponent, largest number approximately 10^{2465}). In such cases, scaling of C1 and C2 (used only in forming the ratio C1/C2) should be added to the code.

If we are only interested in stencils that extend over all the gridpoints $X(0), X(1), \ldots, X(N)$, the following variation saves storage by requiring only a 2-D array for the coefficients. The arithmetic operations (which are the same as in the previous case) are re-ordered in such a way that intermediate results can safely overwrite each other as the calculation proceeds.

```
      SUBROUTINE WEIGHTS1 (XI,X,N,M,C)
C +----------------------------------------------------------------+
C | INPUT PARAMETERS:                                              |
C |   XI  POINT AT WHICH THE APPROXIMATIONS ARE TO BE ACCURATE     |
C |   X   X-COORDINATES FOR GRID POINTS, ARRAY DIMENSIONED X(0:N)  |
C |   N   THE GRID POINTS ARE AT X(0),X(1),...,X(N) (I.E. N+1 IN ALL) |
C |   M   HIGHEST ORDER OF DERIVATIVE TO BE APPROXIMATED           |
C |                                                                |
C | OUTPUT PARAMETER:                                              |
C |   C   WEIGHTS, ARRAY DIMENSIONED C(0:N,0:M).                   |
C |       ON RETURN, THE ELEMENT C(J,K) CONTAINS THE WEIGHT TO BE  |
C |       APPLIED AT X(J) WHEN THE K:TH DERIVATIVE IS APPROXIMATED |
C |       BY A STENCIL EXTENDING OVER X(0),X(1),...,X(N).          |
C +----------------------------------------------------------------+
      IMPLICIT REAL*8 (A-H,O-Z)
      DIMENSION X(0:N),C(0:N,0:M)
      C1    = 1.0D0
      C4    = X(0)-XI
      DO 10 K=0,M
         DO 10 J=0,N
10          C(J,K) = 0.0D0
      C(0,0) = 1.0D0
      DO 50 I=1,N
         MN  = MIN(I,M)
         C2    = 1.0D0
```

```
         C5     = C4
         C4     = X(I)-Z
         DO 40 J=0,I-1
            C3 = X(I)-X(J)
            C2 = C2*C3
            IF (J.EQ.I-1) THEN
               DO 20 K=MN,1,-1
 20               C(I,K) = C1*(K*C(I-1,K-1)-C5*C(I-1,K))/C2
               C(I,0)     = -C1*C5*C(I-1,0)/C2
            ENDIF
            DO 30 K=MN,1,-1
 30            C(J,K) = (C4*C(J,K)-K*C(J,K-1))/C3
 40         C(J,0) = C4*C(J,0)/C3
 50      C1 = C2
      RETURN
      END
```

The following test driver first calls WEIGHTS to obtain all the entries in Table 3.1-2. The output from this part starts with a table of coefficients for the zeroth derivative, that is, the interpolation weights (they are trivial since the approximations are requested at a gridpoint; the weight is unity at that gridpoint and zero at all others – this is optimal at any order). Following this calculation (and printout), the driver calls WEIGHTS1 to again calculate the weights in the special case of stencils extending across all the gridpoints.

```
      PROGRAM TEST
      PARAMETER (M=4,N=8)
      IMPLICIT REAL*8 (A-H,O-Z)
      DIMENSION X(0:N),C(0:N,0:N,0:M),D(0:N,0:M)
      IO = 6
      DO 10 J=0,N
 10      X(J) = J
      CALL WEIGHTS  (0.0D0,X,N,M,C)
      DO 30 K=0,M
         DO 20 I=1,N
 20         WRITE (IO,40)  (C(J,I,K),J=0,I)
 30      WRITE (IO,*)
 40   FORMAT (1X,9F8.3)
      CALL WEIGHTS1 (0.0D0,X,N,M,D)
      DO 50 K=0,M
 50      WRITE (IO,40)  (D(J,K),J=0,N)
      STOP
      END
```

All the data in Table 3.1-1 can similarly be obtained from a single call to SUBROUTINE WEIGHTS by initializing X(0:8) to /0,-1,1,-2,2,-3,3,-4,4/ (and ignoring every second line of the output).

D

Lebesgue constants

The Lebesgue constants Λ_N express how far off a polynomial interpolant can be from a function when compared to the optimal polynomial of the same (or lower) degree. In the maximum norm:

$$\|f-P_N^{\text{interp}}\| \le (1+\Lambda_N)\|f-P_N^{\text{opt}}\|.$$

For any distribution of interpolation nodes, Λ_N will grow at least as $O(\ln N)$. The lower this growth rate, the more stable is the interpolation procedure against spurious oscillations and other errors.

The value of Λ_N depends in a very sensitive way on the node distribution. Although both Chebyshev and Legendre polynomials feature the same node density function $\mu(x) = 1/(\pi\sqrt{1-x^2})$ (see Section 3.3), their Lebesgue constants are quite different:

$$\Lambda_N^{\text{Ch}} = O(\ln N) \;\; \text{versus} \;\; \Lambda_N^{\text{Leg}} = O(\sqrt{N}).$$

For derivations, see Rivlin (1969, p. 90) and Szegö (1959, p. 336), respectively.

Vértesi (1990) shows that Λ_N^{Ch} is remarkably close to the smallest possible for any node distribution:

$$\Lambda_N^{\text{Ch}} = \frac{2}{\pi}\left(\ln N+\gamma+\ln\frac{8}{\pi}\right)+o(1) \;\; \text{versus} \;\; \Lambda_N^{\text{min}} = \frac{2}{\pi}\left(\ln N+\gamma+\ln\frac{4}{\pi}\right)+o(1).$$

Trefethen and Weideman (1991) survey the history of the equi-spaced case and note the lack of a simple derivation for $\Lambda_N^{\text{eq}} = O(2^N/(N\ln N))$.

The independent derivations of Turetskii (1940) and Schönhage (1961) were both quite complex, and received little attention (in part, perhaps, because they were not published in English). Different weaker estimates have been presented many times.

We next outline a very short derivation for the equi-spaced case. One reason for doing this is that the same general approach can be useful in estimating sizes of the weights in many equi-spaced FD formulas.

Derivation of the Lebesgue constant for equi-spaced interpolation

As noted in Section 3.3,

$$\Lambda_N^{\text{eq}} = \max_{x\in[-1,1]} \sum_{k=0}^{N} |F_k(x)|. \tag{D-1}$$

The term $F_k(x)$ is defined in equation (3.3-2), and the nodes are assumed to be equi-spaced over $[-1,1]$; that is, $x_j = -1+2j/N$, $j=0,1,\ldots,N$.

If (in max-norm) $\|f-P_N^{\text{opt}}\| \le \epsilon$, the function values used for the interpolation can at worst be ϵ off from the values of P_N^{opt} at these locations. The right-hand side in (D-1), multiplied by ϵ, represents the maximal change these perturbations can lead to via Lagrange's interpolation formula - assuming every data point is "off" by the largest allowed amount ϵ and the signs are such that the errors accumulate (rather than cancel).

The size of the interval we consider has no bearing on the value of Λ_N^{eq}. To keep the notation simple, we consider $[0,N]$ instead of $[-1,1]$.

Figure 3.3-3(a) strongly suggested (and it is easy to verify) that the maximum is to be found in the first interval $[0,1]$ (and in the last, $[N-1,N]$). From (D-1) and

$$F_k(x) = \frac{x\cdot(x-1)\cdots(x-k+1)\cdot(x-k-1)\cdots(x-N)}{k\cdot(k-1)\cdots1\cdot(-1)\cdots(k-N)},$$

it follows that

$$\begin{aligned}\Lambda_N^{\text{eq}} &= \max_{x\in[0,1]} \sum_{k=0}^{N}\left|\frac{x\cdot(x-1)\cdots(x-k+1)\cdot(x-k-1)\cdots(x-N)}{k\cdot(k-1)\cdots1\cdot(-1)\cdots(k-N)}\right| \\ &= \max_{x\in[0,1]} \sum_{k=0}^{N} \frac{x\cdot(1-x)\cdot(2-x)\cdots(N-x)}{|k-x|\cdot k!\cdot(N-k)!}.\end{aligned}$$

Repeated use of $\Gamma(z+1) = z\Gamma(z)$ gives

$$\frac{\Gamma(N+1-x)}{\Gamma(1-x)} = (N-x)(N-1-x)\cdots(1-x).$$

Thus,

$$\Lambda_N^{\text{eq}} = \max_{x\in[0,1]} \frac{x\Gamma(N+1-x)}{\Gamma(1-x)} \sum_{k=0}^{N} \frac{1}{|k-x|\cdot k!\cdot(N-k)!}.$$

Up to this point, everything has been exact. We now make two approximations that are easily seen to be correct to leading order as $N \to \infty$:

1. $\Gamma(N+1-x) \sim N! \cdot N^{-x}$; and
2. for increasing N, the sum becomes more and more dominated by the terms around $k = N/2$, so we replace $1/|k-x|$ by $2/N$ and factor this out.

This gives

$$\Lambda_N^{\text{eq}} \approx \max_{x \in [0,1]} \frac{xN^{-x}}{\Gamma(1-x)} \cdot \frac{2}{N} \sum_{k=0}^{N} \frac{N!}{k! \cdot (N-k)!}.$$

We proceed by noting that:

3. the sum is the binomial expansion of $(1+1)^N = 2^N$;
4. as $N \to \infty$, the function xN^{-x} develops an increasingly sharp local maximum located at $x = 1/(\ln N)$ and takes there the value $1/(e \cdot \ln N)$;
5. $\Gamma(1-x) \to 1$ as $x \to 0$; and
6. the right-hand side is growing rapidly toward infinity for increasing N (we can ignore the constant 1 on the left-hand side).

Therefore,

$$\Lambda_N^{\text{eq}} \approx \frac{2^{N+1}}{e \cdot N \cdot \ln N} = O\left(\frac{2^N}{N \cdot \ln N}\right).$$

The coarsest of the foregoing estimates was at item 5. The more precise statement $\Gamma(1-x) = 1 + \gamma x + O(x^2)$ would lead directly to the sharper result

$$\Lambda_N^{\text{eq}} \approx \frac{2^{N+1}}{e \cdot N \cdot (\gamma + \ln N)}.$$

E

Potential function estimate for polynomial interpolation error

Defining $\omega(x)=\prod_{j=0}^{N}(x-x_j)$, the functions $F_k(x)$ in (3.3-2) can be written

$$F_k(x)=\frac{\omega(x)}{(x-x_k)\omega'(x_k)}, \quad k=0,1,\ldots,N. \tag{E-1}$$

Both sides represent the unique Nth-degree polynomial that satisfies (3.3-3). The case $j=k$ follows from l'Hôpital's rule.

Since we will consider interpolation at complex locations, we now swap notation and label the interpolation point as z rather than x. However, we continue to denote the nodes as x_k, $k=0,1,\ldots,N$, since these remain located on $[-1,1]$. In what follows we assume that the function $f(z)$ is analytic in some neighborhood of $[-1,1]$.

The remainder term $R_N(z)=f(z)-p_N(z)$ can be written

$$R_N(z)=f(z)-\sum_{k=0}^{N}\frac{\omega(z)f(x_k)}{(z-x_k)\omega'(x_k)}$$ from (3.3-1) and (E-1)

$$=\frac{\omega(z)}{2\pi i}\int_C \frac{f(t)}{\omega(t)(t-z)}\,dt,$$ by calculus of residues: the integrand has simple poles at $t=z$ with residue $f(z)/\omega(z)$, giving $f(z)$, and at $t=x_k$, $k=0,1,\ldots,N$, with residue $f(x_k)/\omega'(x_k)(x_k-z)$, giving the terms in the sum

where the contour C encloses $[-1,1]$ and the interpolation location z, but not any singularity of the function $f(z)$.

The contour integral is known as *Hermite's formula*. Mayers (1966) describes one way to arrive at it in a systematic manner, rather than simply stating and verifying it (as we have done here).

To estimate $R_N(z)$, we first need to estimate $\omega(z)$:

$$|\omega(z)| = \prod_{k=0}^{N} |z-x_k| = \exp\left[\sum_{k=0}^{N} \ln|z-x_k|\right]$$
$$\approx e^{-N\phi(z)}.$$

The estimate follows by noting that

$$\frac{1}{N}\sum_{k=0}^{N} \ln|z-x_k| \to \int_{-1}^{1} \mu(x)\ln|z-x|\,dx = -\phi(z),$$

as defined in (3.4-1) – assuming for the moment that the constant is zero.

We choose any $\epsilon > 0$ and select C as a contour curve for $\phi(z)$ (i.e. for $|\omega(z)|$) such that $\phi(z)_{\text{on}\,C} > \phi(z_0)+\epsilon$ (contour C inside the nearest singularity z_0 of $f(z)$). Then

$$\left|\int_C \frac{f(t)\,dt}{\omega(t)(t-z)}\right| \le \left\{\int_C \left|\frac{f(t)}{t-z}\right| dt\right\} \frac{1}{|\omega(z)|_{\text{on}\,C}} \le c(\epsilon)e^{N(\phi(z_0)+\epsilon)}$$

and

$$|R_N(z)| \le c(\epsilon)e^{N[(\phi(z_0)+\epsilon)-\phi(z)]}.$$

In this estimate, $\phi(z_0)+\epsilon$ cannot be improved to $\phi(z_0)-\epsilon$, because in that case $|R_N(z)| \to 0$ exponentially fast at the location $z = z_0$ (where $f(z)$ is singular) – an impossibility. Therefore

$$|R_N(z)|^{1/N} \to e^{-(\phi(z)-\phi(z_0))} \tag{E-2}$$

as $N \to \infty$.

This argument applies only to values of z inside the particular contour C that was used in the estimate. If z is outside a contour that encloses z_0, then $R_N(z)$ diverges. The divergence rate is again given by (E-2), since it then becomes irrelevant that we no longer pick up the residue at $t = z$.

F

FFT-based implementation of PS methods

Periodic PS methods are almost always implemented with use of the FFT algorithm. For nonperiodic PS methods, direct matrix × vector multiplication is often both fast and convenient. However, in the case of Chebyshev–PS methods, a cosine–FFT approach is also effective. Following a description of the FFT concept in Section F.1, its use for periodic and Chebyshev–PS implementations is described in Sections F.2 and F.3.

In most periodic PS contexts, what is actually needed is not Fourier expansion coefficients but rather a fast way to compute periodic convolutions. FFTs offer one way to do this. In Section F.4, we discuss convolutions and some alternative ways to calculate them effectively. In Section F.5, we find that, at four times the cost of a "basic" FFT, the scope of the algorithm can be greatly broadened. These fractional Fourier transforms apply to many problems of physical and computational interest.

F.1. The FFT algorithm

Given N complex numbers u_j, $j = 0, 1, \ldots, N-1$, the physical-to-Fourier DFT (discrete Fourier transform)

$$\hat{u}_k = \frac{1}{N} \sum_{j=0}^{N-1} u_j e^{-2\pi ikj/N}, \quad k = 0, 1, \ldots, N-1, \tag{F.1-1}$$

produces equally many complex Fourier coefficients $\hat{u}_k$. The Fourier-to-physical DFT

$$u_j = \sum_{k=0}^{N-1} \hat{u}_k e^{2\pi ikj/N}, \quad j = 0, 1, \ldots, N-1, \tag{F.1-2}$$

recovers the original numbers u_j, $j = 0, 1, \ldots, N-1$.

It is often very convenient to write these equations as matrix × vector products. For example, in the case of (F.1-2) we have

$$\begin{bmatrix} 1 & 1 & 1 & 1 & \cdots & 1 \\ 1 & \omega & \omega^2 & \omega^3 & \cdots & \omega^{N-1} \\ 1 & \omega^2 & \omega^4 & \omega^6 & \cdots & \omega^{2N-2} \\ \vdots & \vdots & & & & \\ \vdots & \vdots & & & & \\ 1 & \omega^{N-1} & \cdots & \cdots & \cdots & \omega^{(N-1)^2} \end{bmatrix} \begin{bmatrix} \hat{u}_0 \\ \hat{u}_1 \\ \hat{u}_2 \\ \vdots \\ \vdots \\ \hat{u}_{N-1} \end{bmatrix} = \begin{bmatrix} u_0 \\ u_1 \\ u_2 \\ \vdots \\ \vdots \\ u_{N-1} \end{bmatrix}, \tag{F.1-3}$$

where $\omega = e^{2\pi i/N}$.

The fast Fourier transform (FFT) algorithm amounts to noting that this DFT matrix can be written as a product of a few sparse matrices (thus replacing the full matrix × vector multiplication in (F.1-3) with a few sparse matrix × vector multiplications). This factorization becomes particularly simple and economical when N is a highly composite number, in particular a power of 2.

Table F.1-1 illustrates the key factorization step in the case of $N = 8$. The DFT matrix is seen to split into a product of a sparse matrix, two DFT matrices of half the size, and a permutation matrix. Repetition of this process leads immediately to the top factorization shown in Table F.1-2. FFT factorizations are far from unique. The "self-sorting" Glassman (1970) algorithm (Table F.1-2, bottom) can be obtained by incorporating the permutations into the sparse factors. The FFT code below uses the Cooley–Tukey algorithm, which has the advantage that no extra data vector is needed for intermediate storage during the matrix × vector multiplications. Extensive discussions and references about FFT variations can be found for example in van Loan (1992).

```
C-----------------------------------------------------------------------
      SUBROUTINE FFT (A,B,IS,N,ID)
C-- +-------------------------------------------------------------------+
C-- | A CALL TO FFT REPLACES THE COMPLEX DATA VALUES A(J) + i B(J),     |
C-- | J=0,1,...,N-1 WITH THEIR TRANSFORM                                |
C-- |                                                                   |
C-- |                               2 i ID PI K J / N                   |
C-- |    SUM      (A(K) + i B(K)) e                  , J=0,1,...,N-1    |
C-- |  K=0..N-1                                                         |
C-- |                                                                   |
C-- | INPUT AND OUTPUT PARAMETERS                                       |
C-- |     A    ARRAY A(0:*), REAL PART OF DATA/TRANSFORM                |
C-- |     B    ARRAY B(0:*), IMAGINARY PART OF DATA/TRANSFORM           |
C-- | INPUT PARAMETERS                                                  |
C-- |    IS    SPACING IN MEMORY BETWEEN CONSECUTIVE ELEMENTS IN A AND B |
C-- |          (USE IS=+1 FOR ELEMENTS STORED CONSECUTIVELY)            |
C-- |     N    NUMBER OF DATA VALUES, MUST BE A POWER OF TWO            |
C-- |    ID    USE +1 OR -1 TO SPECIFY DIRECTION OF TRANSFORM           |
C-- |                                                                   |
C-- +-------------------------------------------------------------------+
```

Table F.1-1. *Splitting of DFT matrix into a sparse product*

DFT MATRIX

0	**0**	**0**	**0**	**0**	**0**	**0**	**0**
0	**1**	**2**	**3**	**4**	**5**	**6**	**7**
0	**2**	**4**	**6**	**8**	**10**	**12**	**14**
0	**3**	**6**	**9**	**12**	**15**	**18**	**21**
0	**4**	**8**	**12**	**16**	**20**	**24**	**28**
0	**5**	**10**	**15**	**20**	**25**	**30**	**35**
0	**6**	**12**	**18**	**24**	**30**	**36**	**42**
0	**7**	**14**	**21**	**28**	**35**	**42**	**49**

=

RE-ORDER THE COLUMNS EVEN - ODD

0	**0**	**0**	**0**	**0**	**0**	**0**	**0**
0	**2**	**4**	**6**	**1**	**3**	**5**	**7**
0	**4**	**8**	**12**	**2**	**6**	**10**	**14**
0	**6**	**12**	**18**	**3**	**9**	**15**	**21**
0	**8**	**16**	**24**	**4**	**12**	**20**	**28**
0	**10**	**20**	**30**	**5**	**15**	**25**	**35**
0	**12**	**24**	**36**	**6**	**18**	**30**	**42**
0	**14**	**28**	**42**	**7**	**21**	**35**	**49**

x

1							
		1					
				1			
						1	
	1						
			1				
					1		
							1

⇓ ⇓

REDUCE INDICES

0	**0**	**0**	**0**	**0**	**0**	**0**	**0**
0	**2**	**4**	**6**	**1**	**3**	**5**	**7**
0	**4**	**8**	**12**	**2**	**6**	**10**	**14**
0	**6**	**12**	**18**	**3**	**9**	**15**	**21**
0	**0**	**0**	**0**	**-0**	**-0**	**-0**	**-0**
0	**2**	**4**	**6**	**-1**	**-3**	**-5**	**-7**
0	**4**	**8**	**12**	**-2**	**-6**	**-10**	**-14**
0	**6**	**12**	**18**	**-3**	**-9**	**-15**	**-21**

⇓ ⇓

SPLIT INTO SPARSE MATRIX

0				**0**			
	0				**1**		
		0				**2**	
			0				**3**
0				**-0**			
	0				**-1**		
		0				**-2**	
			0				**-3**

x

2 DFTs OF HALF SIZE

0	**0**	**0**	**0**				
0	**2**	**4**	**6**				
0	**4**	**8**	**12**				
0	**6**	**12**	**18**				
				0	**0**	**0**	**0**
				0	**2**	**4**	**6**
				0	**4**	**8**	**12**
				0	**6**	**12**	**18**

x

PERMUTATION MATRIX

1							
		1					
				1			
						1	
	1						
			1				
					1		
							1

Notes: In this illustration, $N = 8$. Boldface entries $\boldsymbol{k}$ and $-\boldsymbol{k}$ are shorthand for ω^k and $-\omega^k$, respectively. The relations $\omega^{k+N} = \omega^k$ and $\omega^{k+N/2} = -\omega^k$ are used repeatedly.

```
      IMPLICIT REAL*8 (A-H,O-Z)
      DIMENSION A(0:*),B(0:*)
      J = 0
C---   APPLY PERMUTATION MATRIX    --------
      DO 20 I=0,(N-2)*IS,IS
         IF (I.LT.J) THEN
            TR   = A(J)
            A(J) = A(I)
            A(I) = TR
```

Table F.1-2. *Illustration of FFT techniques*

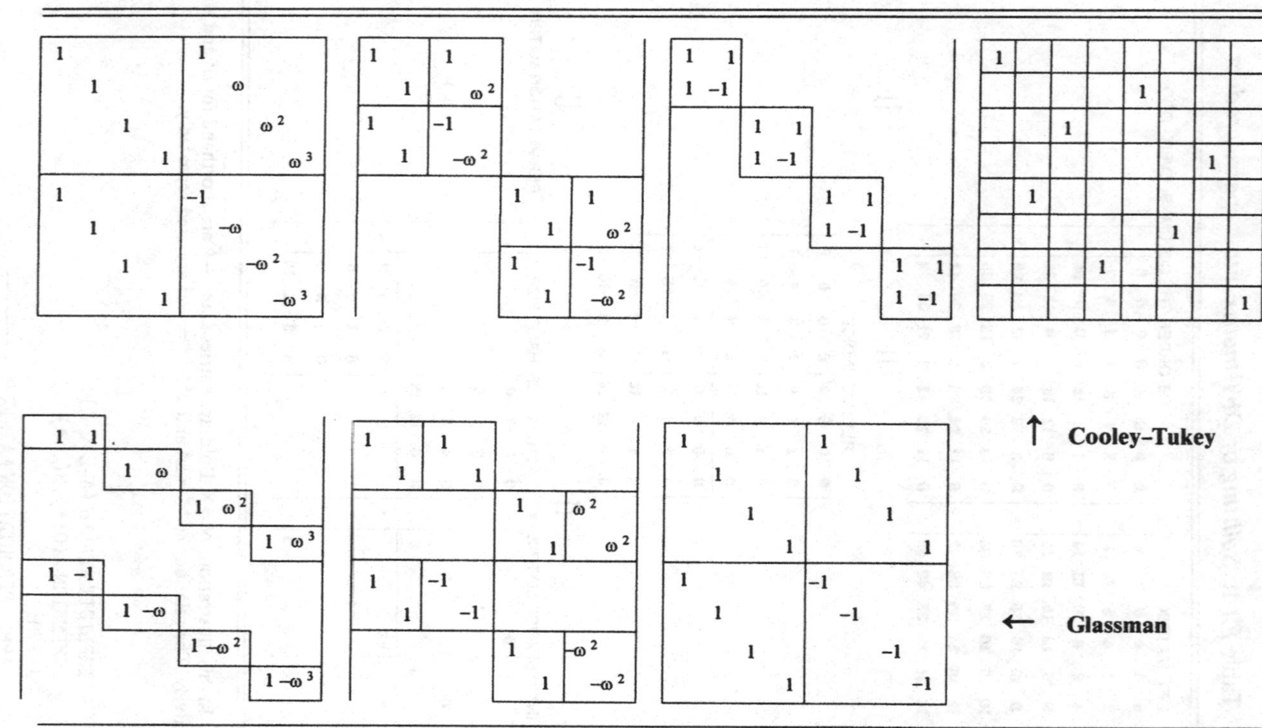

Note: The table illustrates two different sparse factorizations of the 8×8 DFT matrix (as defined in equation F.1-3).

```
            TI   = B(J)
            B(J) = B(I)
            B(I) = TI
         ENDIF
         K = IS*N/2
   10    IF (K.LE.J) THEN
            J = J-K
            K = K/2
            GOTO 10
         ENDIF
   20    J = J+K
C---   PERFORM THE LOG2 N MATRIX-VECTOR MULTIPLICATIONS     ---
      S =  0.0D0
      C = -1.0D0
      L = IS
   30    LH = L
         L  = L+L
         UR = 1.0D0
         UI = 0.0D0
         DO 50 J=0,LH-IS,IS
            DO 40 I=J,(N-1)*IS,L
               IP = I+LH
               TR = A(IP)*UR-B(IP)*UI
               TI = A(IP)*UI+B(IP)*UR
               A(IP) = A(I)-TR
               B(IP) = B(I)-TI
               A(I) = A(I)+TR
   40          B(I) = B(I)+TI
            TI = UR*S+UI*C
            UR = UR*C-UI*S
   50       UI = TI
         S = SQRT (0.5D0*(1.0D0-C))*ID
         C = SQRT (0.5D0*(1.0D0+C))
         IF (L.LT.N*IS) GOTO 30
      RETURN
      END
```

A complex FFT code like this one takes as input two vectors $A(I)$ and $B(I)$, $I=0,1,\ldots,N-1$ (containing real and imaginary parts), and produces two similar output vectors $C(I)$ and $D(I)$ (which might, as in our FFT routine, overwrite the input vectors). The top row of displays in Table F.1-3 illustrate these input and output data vectors. Some special cases arise commonly in applications.

Real data: The output of a complex FFT will in this case feature the structure shown in the second row of Table F.1-3. Special codes that operate only on the numbers within the heavy lines can be made approximately twice as fast as general complex FFT codes. The two easiest ways to handle this situation of real data vectors are as follows.

Table F.1-3. *Complex FFTs –*

Transform type		Structure of input									
Complex		General, complex									
	Re	a_0	a_1	…	…	$a_{N/2-1}$	$a_{N/2}$	$a_{N/2+1}$	…	…	a_{N-1}
	Im	b_0	b_1	…	…	$b_{N/2-1}$	$b_{N/2}$	$b_{N/2+1}$	…	…	b_{N-1}
Real		Real									
	Re	a_0	a_1	…	…	$a_{N/2-1}$	$a_{N/2}$	$a_{N/2+1}$	…	…	a_{N-1}
	Im	0	0	…	…	0	0	0	…	…	0
Cosine		Real, symmetric									
	Re	a_0	a_1	…	…	$a_{N/2-1}$	$a_{N/2}$	$a_{N/2-1}$	…	…	a_1
	Im	0	0	…	…	0	0	0	…	…	0
Sine		Real, antisymmetric									
	Re	0	a_1	…	…	$a_{N/2-1}$	0	$-a_{N/2-1}$	…	…	$-a_1$
	Im	0	0	…	…	0	0	0	…	…	0

Notes: The heavy lines surround data vectors that the different routines require as input and

(1) Place pairs of real data vectors as Re and Im parts in the input of a complex transform (this is very simple and effective – see Section F.2).
(2) Use fast cosine and sine transforms – for example, the routines FCT and FST, which are discussed next.

Symmetric or antisymmetric real data: In these cases, specialized codes (such as FCT and FST) run nearly four times as fast as would a general complex FFT. Both FCT and FST are identical to their inverses; that is, there is no need to specify whether data is moved from physical to Fourier space or in the opposite direction.

general versus special cases

Structure of output	Name of code included
General, complex	FFT
Real part symmetric, imaginary part antisymmetric	—
Real, symmetric	FCT
Imaginary, antisymmetric	FST

General, complex (FFT)

Re	c_0	c_1	…	…	$c_{N/2-1}$	$c_{N/2}$	$c_{N/2+1}$	…	…	c_{N-1}
Im	d_0	d_1	…	…	$d_{N/2-1}$	$d_{N/2}$	$d_{N/2+1}$	…	…	d_{N-1}

Real part symmetric, imaginary part antisymmetric (—)

Re	c_0	c_1	…	…	$c_{N/2-1}$	$c_{N/2}$	$c_{N/2-1}$	…	…	c_1
Im	0	d_1	…	…	$d_{N/2-1}$	0	$-d_{N/2-1}$	…	…	$-d_1$

Real, symmetric (FCT)

Re	c_0	c_1	…	…	$c_{N/2-1}$	$c_{N/2}$	$c_{N/2-1}$	…	…	c_1
Im	0	0	…	…	0	0	0	…	…	0

Imaginary, antisymmetric (FST)

Re	0	0	…	…	0	0	0	…	…	0
Im	0	d_1	…	…	$d_{N/2-1}$	0	$-d_{N/2-1}$	…	…	$-d_1$

provide as output. N is assumed to be even.

The routine FCT can be used for effective implementations of Chebyshev–PS methods (to be discussed in Section F.3). To achieve the most convenient notation for that context, the variable N in FCT (and in FST) denotes the actual length of the input and output data vectors (i.e., what is denoted by $N/2$ in Table F.1-3).

A real vector can easily be split up as the sum of a real symmetric vector and a real antisymmetric vector. These can then be transformed by FCT and FST respectively, giving the two parts of the output that a general real FFT would have produced (cf. again Table F.1-3).

Codes for FCT and FST will now be listed. In both cases, the output vector B may be the same as the input vector A (i.e., the data overwriting that will occur within the routines is safe). Each routine calls the complex routine FFT once (with a vector length of $N/2$ – or a length of $N/4$ in the notation of Table F.1-3).

```
C-----------------------------------------------------------------------
      SUBROUTINE FCT (A,X,N,B)
C-- +-------------------------------------------------------------------+
C-- | A CALL TO FCT PLACES IN B(0:N) THE COSINE TRANSFORM OF THE        |
C-- | VALUES IN A(0:N)                                                  |
C-- |                                                                   |
C-- |      B(J)  =   SUM    C(K)*A(K)*COS(PI*K*J/N) ,  J=0,1,...,N      |
C-- |              K=0..N                                               |
C-- |                                                                   |
C-- |      WHERE  C(K) = 1.0 FOR K=0,N,  C(K) = 2.0 FOR K=1,2,...,N-1   |
C-- |                                                                   |
C-- | INPUT PARAMETERS:                                                 |
C-- |      A   A(0:N)   ARRAY WITH INPUT DATA                           |
C-- |      X   X(0:N)   ARRAY WITH CHEBYSHEV GRID POINT LOCATIONS       |
C-- |          X(J) = -COS(PI*J/N) ,  J=0,1,...,N                       |
C-- |      N   SIZE OF TRANSFORM - MUST BE A POWER OF TWO               |
C-- | OUTPUT PARAMETER                                                  |
C-- |      B   B(0:N)   ARRAY RECEIVING TRANSFORM COEFFICIENTS          |
C-- |          (MAY BE IDENTICAL WITH THE ARRAY A)                      |
C-- |                                                                   |
C-- +-------------------------------------------------------------------+
      IMPLICIT REAL*8 (A-H,O-Z)
      DIMENSION A(0:*),X(0:*),B(0:*)
      N2 = N/2
      A0 = A(N2-1)+A(N2+1)
      A9 = A(1)
      DO 10 I=2,N2-2,2
        A0 = A0+A9+A(N+1-I)
        A1 = A(  I+1)-A9
        A2 = A(N+1-I)-A(N-1-I)
        A3 = A(I)+A(N-I)
        A4 = A(I)-A(N-I)
        A5 = A1-A2
        A6 = A1+A2
        A1 = X(N2-I)
        A2 = X(I)
        A7 = A1*A4+A2*A6
        A8 = A1*A6-A2*A4
        A9 = A(I+1)
        B(I  ) = A3+A7
        B(N-I) = A3-A7
        B(I+1  ) = A8+A5
   10   B(N+1-I) = A8-A5
      B(1) = A(0)-A(N)
      B(0) = A(0)+A(N)
```

```
      B(N2  ) = 2.D0*A(N2)
      B(N2+1) = 2.D0*(A9-A(N2+1))
      CALL FFT (B(0),B(1),2,N2,1)
      A0 = 2.D0*A0
      B(N) = B(0)-A0
      B(0) = B(0)+A0
      DO 20 I=1,N2-1
         A1 = 0.5 D0          *(B(I)+B(N-I))
         A2 = 0.25D0/X(N2+I)*(B(I)-B(N-I))
         B(I  ) = A1+A2
   20    B(N-I) = A1-A2
      RETURN
      END

C-----------------------------------------------------------------------
      SUBROUTINE FST (A,X,N,B)
C-- +-------------------------------------------------------------------+
C-- | A CALL TO FST PLACES IN B(1:N-1) THE SINE TRANSFORM OF THE        |
C-- | VALUES IN A(1:N-1) (AND PUTS B(0)=B(N)=0.D0).                     |
C-- |                                                                   |
C-- |    B(J)  =  2.0 *  SUM    A(K)*SIN(PI*K*J/N) ,  J=1,2,...,N-1      |
C-- |                  K=1..N-1                                         |
C-- |                                                                   |
C-- | INPUT PARAMETERS:                                                 |
C-- |    A   A(0:N)  ARRAY WITH INPUT DATA   ( A(0) AND A(N) IGNORED )  |
C-- |    X   X(0:N)  ARRAY WITH CHEBYSHEV GRID POINT LOCATIONS          |
C-- |        X(J) = -COS(PI*J/N) ,  J=0,1,...,N                         |
C-- |    N   SIZE OF TRANSFORM - MUST BE A POWER OF TWO                 |
C-- | OUTPUT PARAMETER                                                  |
C-- |    B   B(0:N)  ARRAY RECEIVING TRANSFORM COEFFICIENTS             |
C-- |        (MAY BE IDENTICAL WITH THE ARRAY A)                        |
C-- |                                                                   |
C-- +-------------------------------------------------------------------+
      IMPLICIT REAL*8 (A-H,O-Z)
      DIMENSION A(0:*),X(0:*),B(0:*)
      N2 = N/2
      A9 = A(1)
      B(0) = 2.D0*(A9-A(N-1))
      B(1) = 2.D0*(A(N-1)+A9)
      DO 10 I=2,N2-2,2
         A1 = A9       -A(  I+1)
         A2 = A(N-1-I)-A(N+1-I)
         A3 = A(I)+A(N-I)
         A4 = A(I)-A(N-I)
         A5 = A1-A2
         A6 = A1+A2
         A1 = X(N2+I)
         A2 = X(I)
         A7 = A1*A5+A2*A3
         A8 = A1*A3-A2*A5
         A9      = A(I+1)
         B(I  ) = -A6-A7
```

```
         B(N-I) = -A6+A7
         B(I+1  ) = -A4-A8
10       B(N+1-I) = -A8+A4
      A1        = A(N2)
      B(N2  ) =  2.DO*(A(N2+1)-A9)
      B(N2+1) = -2.DO*A1
      CALL FFT (B(0),B(1),2,N2,-1)
      B(N2) = 0.5DO*B(N2)
      DO 20 I=1,N2-1
         A1 = 0.5 DO           *(B(I)-B(N-I))
         A2 = 0.25DO/X(N2+I)*(B(I)+B(N-I))
         B(I  ) = A2-A1
20       B(N-I) = A2+A1
      B(0) = 0.DO
      B(N) = 0.DO
      RETURN
      END
```

F.2. Periodic PS implementation using the FFT algorithm

The exponentials, and therefore the sum, in (F.1-1) are not changed if any multiple of N is added to or subtracted from k. We can therefore interpret $\hat{u}_{N-j}$ as $\hat{u}_{-j}$, $j = 1, 2, \ldots, N/2-1$ (and write $\hat{u}_{N/2}$ as $\hat{u}_{\pm N/2}$, viewing it as being made up of equal amounts of modes $N/2$ and $-N/2$, the *Nyquist* or *reflection* frequency). This interpretation of the mode numbers turns out to be necessary when determining the interpolating trigonometric polynomial of lowest degree, finding its derivative, and so on. Otherwise, the trigonometric polynomial takes complex values in between the real data points and so its derivatives become nonsensical.

With the modes $k = 0, 1, \ldots, N-1$ interpreted as modes $k = 0, 1, \ldots, N/2-1$, $\pm N/2, -N/2+1, \ldots, -1$, real data yields a real trigonometric interpolant. This can be seen as follows.

(1) With u_j, $j = 0, 1, \ldots, N-1$ real, taking the complex conjugate of $\hat{u}_k = (1/N)\sum_{j=0}^{N-1} u_j e^{-2\pi ikj/N}$ will have the same effect as changing k for $-k$; that is,

$$\overline{\hat{u}_k} = \hat{u}_{-k} \quad \text{for all } k$$

(in particular, $\hat{u}_0$ and $\hat{u}_{\pm N/2}$ are therefore real).

(2) For all values of j (also non-integer), we have

$$u_j = \sum_{k=-N/2+1}^{N/2-1} \hat{u}_k e^{2\pi ikj/N} + \hat{u}_{\pm N/2}\left(\frac{e^{\pi ij}+e^{-\pi ij}}{2}\right).$$

This expression is real, since

$$\overline{u_j} = \overline{\sum_{k=-N/2+1}^{N/2-1} \hat{u}_k e^{2\pi ikj/N} + \hat{u}_{\pm N/2}\cos \pi j} =$$

$$= \sum_{k=N/2+1}^{N/2-1} \hat{u}_{-k} e^{2\pi i(-k)j/N} + \hat{u}_{\pm N/2} \cos \pi j$$

$$= \sum_{k=-N/2+1}^{N/2-1} \hat{u}_k e^{2\pi ikj/N} + \hat{u}_{\pm N/2} \cos \pi j = u_j.$$

Requiring the interpolating trigonometric polynomial to take the values u_j at locations $x_j = -1 + jh$, $j = 0, 1, \ldots, N-1$, $h = 2/N$, leads to the following closed form (now expressed not in j, but in $x \in [-1, 1]$):

$$u(x) = \left\{ \sum_{k=-N/2+1}^{N/2-1} \hat{u}_k e^{\pi ik(x+1)} \right\} + \hat{u}_{\pm N/2} \cos \frac{\pi N}{2}(x+1).$$

Its first derivative is

$$u'(x) = \left\{ \sum_{k=-N/2+1}^{N/2-1} \pi ik\hat{u}_k e^{\pi i(x+1)} \right\} - \frac{\pi N}{2} \hat{u}_{\pm N/2} \sin \frac{\pi N}{2}(x+1).$$

At gridpoints x_j, the modes 0 and $\pm N/2$ do not contribute anything to the derivative. This leads to the following, easily implemented, three-step procedure for obtaining values for the first derivative by the periodic PS method.

1. Perform a complex FFT (according to equation (F.1-1)) on the data values at the gridpoints.
2. Multiply the output elements from this FFT,

$$\hat{u}_0 \ \hat{u}_1 \ \hat{u}_2 \ \ldots \ \hat{u}_{N/2-1} \ \hat{u}_{N/2} \ \hat{u}_{N/2+1} \ \ldots \ \hat{u}_{N/2-2} \ \hat{u}_{N-1},$$

by

$$0 \ \pi i \ 2\pi i \ \ldots \ (N/2-1)\pi i \ 0 \ -(N/2-1)\pi i \ \ldots \ -2\pi i \ -\pi i.$$

3. Perform a complex FFT (according to F.1-2) on these numbers to obtain the PS derivative approximations at the gridpoints.

If the FFT code does not incorporate the $1/N$ factor of (F.1-1), then the constants in step 2 can all be divided by N. If the spatial period is $[a, b]$ instead of $[-1, 1]$, they should all be multiplied by $2/(b-a)$.

The procedure just described uses twice as many arithmetic operations as necessary. As noted in Section F.1, there are two ways to avoid this inefficiency: calculate two real derivatives at the same time; or use real FFTs (or FCTs and FSTs). The first alternative is by far the simplest. Although complex arithmetic is used throughout, the procedure just described gives (as one should expect) a purely real derivative from purely real data. Because the algorithm is linear, we can initialize the imaginary parts with a second instance of a real data vector. Its PS derivative will then – without the need to perform any additional operations – emerge in the imaginary part of the (now complex) derivative. This technique of differentiating two real data vectors at the same time applies immediately if there is more than one dependent function or more than one space variable.

Subroutine	*Function*
TOFOUR	Transform two independent real vectors to Fourier space
DIFFFOUR	Differentiate in Fourier space
FROMFOUR	Return to physical space with pointwise derivatives for the two data vectors

Parameters (same for all three codes)

Input and output:

A, B	A(0:N-1), B(0:N-1) Two separate real data vectors (or Fourier representation of combined data vectors)

Input:

N	Size of data vectors – *must be a power of* 2

Notes

- The calling sequence

```
      CALL TOFOUR (A,B,N)
      DO 10 L=1,M
10       CALL DIFFFOUR (A,B,N)
      CALL FROMFOUR (A,B,N)
```

  replaces A and B with pointwise values for the Mth derivative, $M \geq 0$.
- If lower derivatives (than the Mth) are also needed, only CALL FROMFOUR needs to be repeated.

```
C-----------------------------------------------------------------
      SUBROUTINE TOFOUR (A,B,N)
      IMPLICIT REAL*8 (A-H,O-Z)
      DIMENSION A(0:N-1),B(0:N-1)
      CALL FFT (A,B,1,N,-1)
      A1 = 1.D0/N
      DO 10 I=0,N-1
         A(I) = A(I)*A1
10       B(I) = B(I)*A1
      RETURN
      END

C-----------------------------------------------------------------
      SUBROUTINE DIFFFOUR (A,B,N)
      IMPLICIT REAL*8 (A-H,O-Z)
```

```
      DIMENSION A(0:N-1),B(0:N-1)
      PI = 4.D0*ATAN(1.D0)
      N2 = N/2
      A(0) = 0.D0
      B(0) = 0.D0
      A(N2) = 0.D0
      B(N2) = 0.D0
      DO 10 I=1,N2-1
         J = N-I
         F = I*PI
         A1 = A(I)*F
         A(I) = -B(I)*F
         B(I) = A1
         A1 = A(J)*F
         A(J) = B(J)*F
10       B(J) = -A1
      RETURN
      END

C-----------------------------------------------------------------
      SUBROUTINE FROMFOUR (A,B,N)
      IMPLICIT REAL*8 (A-H,O-Z)
      DIMENSION A(0:N-1),B(0:N-1)
      CALL FFT (A,B,1,N,+1)
      RETURN
      END
```

F.3. Chebyshev–PS implementation using the FFT algorithm

For the Chebyshev method, the data points are not equi-spaced but instead are at locations obtained through

```
      PI = 4.D0*ATAN(1.D0)
      DO 10 I=0,N
10       X(I) = -COS(PI*I/N)
```

The three codes for the Chebyshev case differ from their Fourier counterparts in a few ways:

- they apply to one real data set at a time;
- as input, TOCHEB and FROMCHEB also require the vector $X(I)$ with the Chebyshev node locations (as a source for trigonometric constants – it is unnecessary to recalculate these repeatedly); and
- for all three routines, a separate data output vector is provided. However, it may be set identical with the input vector; that is, output may overwrite the input if desired.

In common with their Fourier counterparts and with the FFT, FCT, and FST routines, the Chebyshev codes:

- use no auxiliary work areas;
- have no upper limit on N (although a lower limit of $N = 4$ is caused by DIFFCHEB); and
- do not call any standard library function except as follows:
  - SQRT is called $2 \log_2 N$ times in FFT;
  - ATAN is called to obtain PI in DIFFFOUR.

  It would be faster to substitute the numerical value into the code, but less portable in cases of computers with unusually high accuracy in double precision.

Subroutine	*Function*
TOCHEB	Applied to a vector containing function values at the $N+1$ Chebyshev node locations $X(I)$, $I=0,1,\ldots,N$, TOCHEB returns the expansion coefficients for the degree-N Chebyshev polynomial that interpolates the data
DIFFCHEB	Manipulates the Chebyshev expansion coefficients corresponding to analytic differentiation
FROMCHEB	Returns from Chebyshev expansion coefficients to pointwise function values at the Chebyshev node locations

Parameters (same for all three codes, apart from the exception noted)

Input:

A	A(0:N) Input data vector
X	X(0:N) Array with Chebyshev node locations (not a parameter for DIFFCHEB)
N	Size of data vectors – *must be a power of* 2

Output:

B	B(0:N) Output data vector (may coincide with A)

```
C-------------------------------------------------------------------
      SUBROUTINE TOCHEB (A,X,N,B)
      IMPLICIT REAL*8 (A-H,O-Z)
      DIMENSION A(0:N),X(0:N),B(0:N)
      CALL FCT (A,X,N,B)
      B1 = 0.5D0/N
      B(0) = B(0)*B1
      B(N) = B(N)*B1
      B1 = 2.D0*B1
      DO 10 I=1,N-1
         B1 = -B1
   10    B(I) = B(I)*B1
      RETURN
      END
```

```
C-----------------------------------------------------------------
      SUBROUTINE DIFFCHEB (A,N,B)
      IMPLICIT REAL*8 (A-H,O-Z)
      DIMENSION A(0:N),B(0:N)
      A1 = A(N)
      A2 = A(N-1)
      B(N) = 0.D0
      B(N-1) = 2.D0*N*A1
      A1 = A2
      A2 = A(N-2)
      B(N-2) = 2.D0*(N-1)*A1
      DO 10 I=N-2,2,-1
         A1 = A2
         A2 = A(I-1)
10       B(I-1) = B(I+1)+2.D0*I*A1
      B(0) = 0.5D0*B(2)+A2
      RETURN
      END

C-----------------------------------------------------------------
      SUBROUTINE FROMCHEB (A,X,N,B)
      IMPLICIT REAL*8 (A-H,O-Z)
      DIMENSION A(0:N),X(0:N),B(0:N)
      B(0) = A(0)
      A1 = 0.5D0
      DO 10 I=1,N-1
         A1 = -A1
10       B(I) = A1*A(I)
      B(N) = A(N)
      CALL FCT (B,X,N,B)
      RETURN
      END
```

Figure F.3-1 compares the computer times for direct DM matrix × vector multiplication against performing the Chebyshev–PS differentiation through

```
CALL TOCHEB   (A,X,N,B)
CALL DIFFCHEB (B,N,B)
CALL FROMCHEB (B,X,N,B)
```

for different N and on two different computers (IBM-compatible PC with 486DX/33 processor and a Cray X-MP).

The codes are quite efficient in all but one of the four cases marked by heavy lines in Figure F.3-1. The exception is the FCT case on the Cray X-MP – the code fails to make good use of the vector hardware of this computer. The dashed line shows the speed improvement that results if the Cray version of IMSL's fast cosine transform routine is substituted for our subroutine FCT (when called from TOCHEB and FROMCHEB).

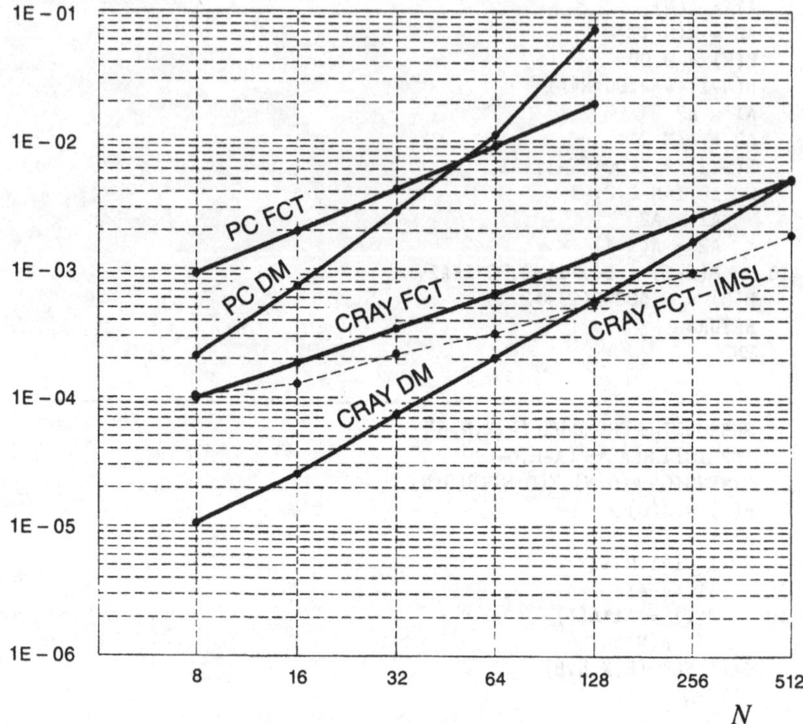

Figure F.3-1. Times required to calculate one set of first derivatives in the Chebyshev–PS method using the DM and FCT approaches. Results are given for an IBM-compatible PC (with 486DX/33 processor) and a Cray X-MP. The matrix × vector multiplication in the DM cases was implemented in the obvious way. In the case labeled CRAY FCT – IMSL, TOCHEB and FROMCHEB did not call our subroutine FCT but rather a functionally equivalent routine from the Cray edition of the IMSL scientific subroutine library. All calculations were performed to 64-bit precision.

For most high-performance computer systems, the manufacturers supply libraries with highly optimized codes for FFTs and other linear algebra tasks such as matrix × vector multiplication. In the particular case of massively parallel machines, such libraries greatly simplify the task of developing effective PS codes.

F.4. Calculation of discrete convolutions

A major application of FFTs is to calculate periodic convolutions.

Discrete convolution theorem (DCT). *If three vectors*

$$[x_0, x_1, \ldots, x_{N-1}],\ [y_0, y_1, \ldots, y_{N-1}],\ [z_0, z_1, \ldots, z_{N-1}] \tag{F.4-1}$$

satisfy

$$\begin{bmatrix} z_0 & z_{N-1} & z_{N-2} & \ddots & z_1 \\ z_1 & z_0 & z_{N-1} & \ddots & z_2 \\ \ddots & \ddots & \ddots & \ddots & \ddots \\ \ddots & \ddots & \ddots & \ddots & \ddots \\ z_{N-1} & z_{N-2} & \ddots & z_1 & z_0 \end{bmatrix} \times \begin{bmatrix} x_0 \\ x_1 \\ \vdots \\ \vdots \\ x_{N-1} \end{bmatrix} = \begin{bmatrix} y_0 \\ y_1 \\ \vdots \\ \vdots \\ y_{N-1} \end{bmatrix}, \tag{F.4-2}$$

then their discrete Fourier coefficients (defined through equation (F.1-1)) *will satisfy*

$$\begin{bmatrix} \hat{z}_0 & & & \\ & \hat{z}_1 & & \\ & & \ddots & \\ & & & \ddots & \\ & & & & \hat{z}_{N-1} \end{bmatrix} \times \begin{bmatrix} \hat{x}_0 \\ \hat{x}_1 \\ \vdots \\ \vdots \\ \hat{x}_{N-1} \end{bmatrix} = \frac{1}{N} \begin{bmatrix} \hat{y}_0 \\ \hat{y}_1 \\ \vdots \\ \vdots \\ \hat{y}_{N-1} \end{bmatrix}.$$

Together with the FFT algorithm, the DCT offers a very fast way to calculate any one of the vectors in (F.4-1) if the other two are provided. This fast convolution procedure has numerous applications. In the context of this book, the application of any FD stencil $[z_0, z_1, \ldots, z_{N-1}]$ to periodic data $[x_0, x_1, \ldots, x_{N-1}]$ amounts to a convolution of the form (F.4-2). Conversely, the equivalent FD stencil to the Fourier–PS method for approximating the first derivative is found by calculating $[z_0, z_1, \ldots, z_{N-1}]$ from

$$[\hat{z}_0, \hat{z}_1, \hat{z}_2, \ldots, \hat{z}_{N-1}] = \frac{i\pi}{N} \times \left[0, 1, 2, \ldots, \frac{N}{2}-1, 0, -\frac{N}{2}+1, \ldots, -2, -1\right].$$

For higher derivatives, we simply need to raise each element in the vector shown here to the corresponding power. For even derivatives, we use $N/2$ as the central element instead of zero. It has here been assumed that N is even and that the physical period is $[-1, 1]$.

Other applications of the DCT include the following.

- *Signal and image processing.* Many signal distortions or image errors (motion blur, bad focusing, etc.) amount to convolutions of the ideal data with some "smearing" function. Deconvolutions with such functions can enhance signals and images.

- *Fast multiplication of polynomials.* The coefficients for the product of

$$p_1(x) = a_0 + a_1x + a_2x^2 + a_3x^3 + \cdots + a_Nx^N$$

and

$$p_2(x) = b_0 + b_1x + b_2x^2 + b_3x^3 + \cdots + b_Nx^N$$

can be found by convolving $[0, 0, \ldots, 0, a_0, a_1, \ldots, a_N]$ with $[0, 0, \ldots, 0, b_0, b_1, \ldots, b_N]$ (where at least as many zeros are needed as there are coefficients).

- By the same technique, very high-precision numbers that are partitioned in segments can be multiplied rapidly. Combined with Newton's method, the same holds true for divisions, square roots, and so forth.

Applying the Fourier–PS method according to the introduction in Chapter 2 amounted to carrying out three steps:

(1) take the FFT of the data;
(2) multiply the discrete Fourier coefficients with their wave numbers (for the first derivative); and
(3) transform back to physical space.

In Chapter 3, the Fourier–PS method was introduced as the limit of infinite order FD schemes, resulting in a periodic convolution. If this convolution is carried out through the DCT, the very same three steps arise again. One might ask if efficient periodic PS implementations must go through these three steps, or if the required convolutions can be performed equally (or even more) efficiently without invoking any trigonometric functions. The answer is that FFT-based methods do indeed seem to be the most convenient way to proceed, but there are two noteworthy alternatives.

Fast number-theoretic transforms

The three circumstances enabling FFT-based fast convolutions were that:

- the DFT matrix in (F.1-3) has an inverse that can be obtained by changing the sign of all the exponents;
- the DFT matrix can be factored, as shown in Table F.1-2; and
- the DCT applies.

For these three results to hold, the only critical requirement is that ω (in equation (F.1-3)) be a "primitive Nth root of unity"; that is, $\omega^k = 1$ for

$k = 0, \pm N, \pm 2N, \ldots$ (and for no other values). The choice $\omega = e^{2\pi i/N}$ is the natural one among the complex numbers. However, if the rules for the basic arithmetic operations are modified, more possibilities might arise. For example, $\omega = 4$ is a primitive root of unity for $N = 4$ if all arithmetic is performed as integers modulo $M = 17$:

$$\omega^0 = 1,\ \omega^1 = 4,\ \omega^2 = 16 \equiv -1,\ \omega^3 = 64 \equiv -4,\ \text{and}\ \omega^4 = 256 \equiv 1.$$

By choosing M sufficiently larger than the elements in the integer vectors to be convolved, the resulting convolution will contain no traces of the unconventional arithmetic used internally during the computation. In applications that do not require floating point calculations (such as extremely high-precision arithmetic or many types of signal processing) this idea opens up attractive possibilities for fast convolutions, as indicated by the following features:

- no need for floating point hardware;
- with suitable choices of M and ω (few binary bits), the transforms do not even require an integer multiplier – just a shifter and an adder;
- no need to calculate and store constants (such as the trigonometric constants in the FFT); and
- the convolutions are obtained entirely free from all rounding errors.

Number-theoretic transforms are surveyed by Agarwal and Burrus (1975) and by Bhattacharya and Agarwal (1984). However, for the floating point–confined numerical analyst, these transforms do not appear practical.

Discrete convolutions with minimal number of multiplications

Winograd (1978) succeeded in determining convolution algorithms with the least possible number of multiplications for any value of N, including prime numbers (where most traditional FFT factorizations fail). These convolution algorithms can be used either directly or as a device to create FFT algorithms that again will use the minimum number of multiplications (about 20% of the typical FFT count; the number of additions is roughly the same). The main reason these algorithms by Winograd are not routinely employed is their complexity (in comparison with the Cooley–Tukey or other FFT algorithms).

Example. Perform the matrix × vector multiplication in (F.4-2) using the minimal number of multiplications in the case of $N = 6$.

The y vector emerges after the following steps:

$$s_1 = x_0 + x_3,\quad s_2 = x_0 - x_3,\quad s_3 = x_1 + x_4,\quad s_4 = x_4 - x_1,$$
$$s_5 = x_2 + x_5,\quad s_6 = x_2 - x_5,\quad s_7 = s_1 + s_3,\quad s_8 = s_7 + s_5,$$
$$s_9 = s_2 + s_4,\quad s_{10} = s_9 + s_6,\quad s_{11} = s_1 - s_3,\quad s_{12} = s_3 - s_5,$$
$$s_{13} = s_5 - s_1,\quad s_{14} = s_2 - s_4,\quad s_{15} = s_4 - s_6,\quad s_{16} = s_6 - s_2;$$

$$m_1 = \frac{z_0+z_1+z_2+z_3+z_4+z_5}{6} s_8,\quad m_2 = \frac{-z_0-z_1+2z_2-z_3-z_4+2z_5}{6} s_{11},$$
$$m_3 = \frac{-2z_0+z_1+z_2-2z_3+z_4+z_5}{6} s_{12},\quad m_4 = \frac{z_0-2z_1+z_2+z_3-2z_4+z_5}{6} s_{13},$$
$$m_5 = \frac{-z_0+z_1-z_2+z_3-z_4+z_5}{6} s_{10},\quad m_6 = \frac{z_0-z_1-2z_2-z_3+z_4+2z_5}{6} s_{14},$$
$$m_7 = \frac{2z_0+z_1-z_2-2z_3-z_4+z_5}{6} s_{15},\quad m_8 = \frac{-z_0-2z_1-z_2+z_3+2z_4+z_5}{6} s_{16};$$

$$s_{17} = m_1 + m_2,\quad s_{18} = s_{17} + m_3,\quad s_{19} = m_1 - m_3,\quad s_{20} = s_{19} + m_4,$$
$$s_{21} = m_1 - m_2,\quad s_{22} = s_{21} - m_4,\quad s_{23} = m_5 + m_6,\quad s_{24} = s_{23} + m_7,$$
$$s_{25} = m_5 - m_7,\quad s_{26} = s_{25} + m_8,\quad s_{27} = m_5 - m_6,\quad s_{28} = s_{27} - m_8.$$

Finally,

$$y_0 = s_{22} - s_{28},\quad y_1 = s_{20} + s_{26},\quad y_2 = s_{18} - s_{24},$$
$$y_3 = s_{22} + s_{28},\quad y_4 = s_{20} - s_{26},\quad y_5 = s_{18} + s_{24}.$$

In typical applications, repeated convolutions are performed with changes only in the x vector. Those parts of the expressions in m_1 to m_8 that involve the z elements would then be precomputed (these expressions can also be split in separate steps using fewer operations, but are given here in their most readable form). If the divisions by 6 (multiplications by 1/6) are left out, the y vector would emerge scaled up by a factor of 6.

F.5. Fractional Fourier transform

The FFT algorithm is extremely effective in calculating the matrix × vector product (F.1-3), and its inverse, when all input values are given and all output values are desired. Direct matrix × vector multiplication is normally much slower, but it does allow cost savings in some frequently occurring situations, as when:

(1) long sections of the input vector contain only zeros;

For example, soliton-type solutions to nonlinear wave equations may be nonzero (to machine precision) over only small sections of a long space domain.

(2) only some of the elements of the output vector are needed.

A large DFT matrix can provide a very accurate trapezoidal-rule approximation of the continuous Fourier integral $f(x) = (1/2\pi)\int_{-\infty}^{\infty} \hat{f}(\omega)e^{i\omega x}\,d\omega$ (or its inverse). Equation (F.1-3) imposes the same uniform discretization levels for $f(x)$ and $\hat{f}(\omega)$. These two functions can have very different characters. For example, we may want to evaluate $f(x)$ at a dense set of points over a small interval in some case where $\hat{f}(\omega)$ is best represented by a sparse set of values over a wide frequency range.

The fractional Fourier transform (FRFT) combines the best of both worlds. To calculate how any m (equi-spaced, e.g., contiguous) entries of the output vector depend on m (also equi-spaced) elements of the input vector, the cost becomes approximately four times that of a size-m FFT (independently of N, the size of the full DFT matrix).

The key observation (Bluestein 1970) is that any sum of the form

$$G_{n,\alpha}(k) = \sum_{j=0}^{m-1} x_j e^{-2\pi i(k+n)j\alpha}, \quad k = 0, 1, \ldots, m-1, \tag{F.5-1}$$
$$n, \alpha \text{ arbitrary constants,}$$

can be reformulated as a (nonperiodic) convolution of size m. As noted in Section F.4, it can then be evaluated efficiently by a periodic convolution of twice the size (or larger, should that be faster or easier). Equation (F.5-1) amounts to a major generalization of (F.1-1) – the latter is recovered as the special case of $n = 0$ and $\alpha = 1/m$.

Noting that $2(k+n)j = j^2 + (k+n)^2 - (k+n-j)^2$, we obtain

$$G_{n,\alpha}(k) = e^{-i\pi(k+n)^2\alpha} \sum_{j=0}^{m-1} (x_j e^{-\pi i j^2\alpha})(e^{\pi i(k+n-j)^2\alpha}).$$

This sum can be written as $\sum_{j=0}^{m-1} y_j z_{k-j}$; i.e., it takes the form of a convolution. If several x vectors are to be used, all exponentials and the transform of the z vector should be precomputed.

Applications of the FRFT extend well beyond the two examples listed at the beginning of this section. Bailey and Swarztrauber (1991, 1994) point out several more, including:

(3) computing DFTs of sequences with arbitrary length N (e.g. a prime number, or not a product of small factors – the FFT routines in Section F.1 required N to be a power of 2);
(4) numerical computation of Laplace transforms (by selecting α to be complex);
(5) analyzing sequences with non-integer periodic components (noting that α need not be a rational number) – for example, an analysis of

seasonal variations from daily measurements (with a year : day ratio of 365.2422...); and

(6) detecting lines in images, and detecting signals with linearly drifting frequencies.

G

Stability domains for some ODE solvers

The primary purpose of this appendix is to illustrate the stability domains of a few classes of common ODE solvers. These diagrams are helpful in selecting ODE packages that are appropriate for time stepping different types of PDEs. Although a few additional comments are made regarding some of the methods, specialized books on ODE solvers (such as those mentioned in Section 4.5) should be consulted for a broader background.

In the context of MOL, we consider the problem of numerically determining a (vector-valued) function $u(t)$, $t>0$, by time stepping of the initial value problem

$$\frac{du}{dt} = f(u,t), \quad u(0) \text{ given.}$$

To conform with conventions for ODEs, we let k denote the size of the time steps with subscripts specifying the sequence number of a time level.

G.1. Runge–Kutta methods

a. Explicit RK methods

The best-known RK method is the following $s=4$–stage scheme, accurate to order $p=4$:

$$\begin{aligned}
d^{(1)} &= kf(u_n, t_n),\\
d^{(2)} &= kf(u_n+\tfrac{1}{2}d^{(1)}, t_n+k/2),\\
d^{(3)} &= kf(u_n+\tfrac{1}{2}d^{(2)}, t_n+k/2),\\
d^{(4)} &= kf(u_n+d^{(3)}, t_n+k);\\
u_{n+1} &= u_n+\tfrac{1}{6}[d^{(1)}+2d^{(2)}+2d^{(3)}+d^{(4)}].
\end{aligned} \tag{G.1-1}$$

Table G.1-1. *Two explicit RK schemes of order 4*

$$\begin{array}{c|cccc} 0 & & & & \\ \frac{1}{2} & \frac{1}{2} & & & \\ \frac{1}{2} & 0 & \frac{1}{2} & & \\ 1 & 0 & 0 & 1 & \\ \hline & \frac{1}{6} & \frac{2}{6} & \frac{2}{6} & \frac{1}{6} \end{array} \qquad \begin{array}{c|cccc} 0 & & & & \\ \frac{1}{3} & \frac{1}{3} & & & \\ \frac{2}{3} & -\frac{1}{3} & 1 & & \\ 1 & 1 & -1 & 1 & \\ \hline & \frac{1}{8} & \frac{3}{8} & \frac{3}{8} & \frac{1}{8} \end{array}$$

Table G.1-2. *Examples of explicit RK schemes of orders 2 and 3*

$$\begin{array}{c|cc} 0 & & \\ \frac{1}{2} & \frac{1}{2} & \\ \hline & 0 & 1 \end{array} \qquad \begin{array}{c|ccc} 0 & & & \\ \frac{1}{3} & \frac{1}{3} & & \\ \frac{2}{3} & 0 & \frac{2}{3} & \\ \hline & \frac{1}{4} & 0 & \frac{3}{4} \end{array}$$

Another $s=4$, $p=4$–method (also due to Kutta) has a somewhat lower leading error term:

$$\begin{aligned} d^{(1)} &= kf(u_n, t_n), \\ d^{(2)} &= kf(u_n + \tfrac{1}{3}d^{(1)}, t_n + k/3), \\ d^{(3)} &= kf(u_n - \tfrac{1}{3}d^{(1)} + d^{(2)}, t_n + 2k/3), \\ d^{(4)} &= kf(u_n + d^{(1)} - d^{(2)} + d^{(3)}, t_n + k); \\ u_{n+1} &= u_n + \tfrac{1}{8}[d^{(1)} + 3d^{(2)} + 3d^{(3)} + d^{(4)}]. \end{aligned}$$

RK schemes are often written compactly as *Butcher arrays*. For the two cases shown here, these arrays are displayed in Table G.1-1. Table G.1-2 shows two explicit RK schemes of orders 2 and 3, respectively.

For different p (order of accuracy), the lowest possible values of s (number of stages) are shown in the following tabulation.

p	1	2	3	4	5	6	7	8	9	10	...
s	1	2	3	4	6	7	9	11	12–17	13–17	...
		$s=p$					$s>p$		$s>p$; exact values unknown		

To estimate local errors (in order to adjust the step size), it is convenient to use special RK schemes that are designed to produce two independent approximations based on the same internal stages. A major advantage of the MOL approach is that such matters are taken care of by ODE packages, and need not concern the user. These packages commonly use methods of orders $p > 4$ (with more complicated Butcher arrays than those shown here).

Although there are many different p-stage methods of order p ($p \leq 4$), their stability domains depend on p only. For $p = 3$ and $p = 4$, Figure G.1-1 shows that sections near the origin of both the imaginary and negative real axes are covered. For $p \geq 5$, a couple of complications arise:

- the stability domains become method-dependent; and
- the order of accuracy may be lower for systems of ODEs than for scalar ODEs.

Neither of these is particularly serious; see Lambert (1991) for details.

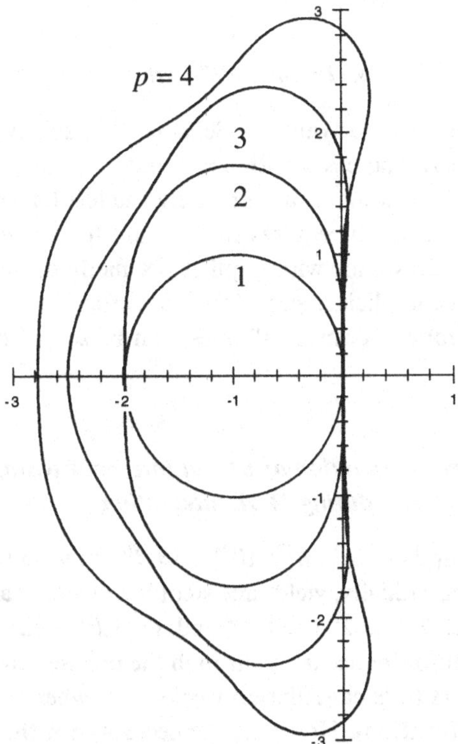

Figure G.1-1. Stability domains for explicit Runge–Kutta methods of orders $p \leq 4$.

Table G.1-3. *Implicit $p=2s$ RK schemes for orders $p=4$ and $p=6$*

Hammer–Hollingsworth			Kuntzmann–Butcher			
			$\frac{1}{2}-\frac{\sqrt{15}}{10}$	$\frac{5}{36}$	$\frac{2}{9}-\frac{\sqrt{15}}{15}$	$\frac{5}{36}-\frac{\sqrt{15}}{30}$
$\frac{1}{2}-\frac{\sqrt{3}}{6}$	$\frac{1}{4}$	$\frac{1}{4}-\frac{\sqrt{3}}{6}$	$\frac{1}{2}$	$\frac{5}{36}+\frac{\sqrt{15}}{24}$	$\frac{2}{9}$	$\frac{5}{36}-\frac{\sqrt{15}}{24}$
$\frac{1}{2}+\frac{\sqrt{3}}{6}$	$\frac{1}{4}+\frac{\sqrt{3}}{6}$	$\frac{1}{4}$	$\frac{1}{2}+\frac{\sqrt{15}}{10}$	$\frac{5}{36}+\frac{\sqrt{15}}{30}$	$\frac{2}{9}+\frac{\sqrt{15}}{15}$	$\frac{5}{36}$
	$\frac{1}{2}$	$\frac{1}{2}$		$\frac{5}{18}$	$\frac{4}{9}$	$\frac{5}{18}$

Runge–Kutta methods are well suited for integrating most convection-dominated problems that are discretized in space by either Fourier– or Chebyshev–PS approximations. For purely convective problems, the stability conditions become $kN <$ const. and $kN^2 <$ const., respectively. The much higher spatial than temporal accuracy will often, even in the latter case, render the time step limited by accuracy – not stability – constraints.

b. *Implicit RK methods*

Again, many methods are available. Schemes that achieve $p = 2s$ can be found for all $s \geq 1$. The cases with $s = 2$ and $s = 3$ are shown in Table G.1-3. Their stability domains are precisely the left half-plane (a special case of A-stability, which requires the domain to *include* the left half-plane). The main drawback with implicit RK methods lies in the cost of solving the coupled implicit stages. However, certain implicit RK methods (e.g. "singly diagonally dominant" ones) can be solved quite effectively by iterations.

c. *Implementation of BCs at internal RK stages during MOL integration*

The tentative updates $d^{(1)}$, $d^{(2)}$, $d^{(3)}$, and $d^{(4)}$ to u_n in the RK4 scheme (G.1-1), if used individually, yield time stepping methods accurate only to (resp.) orders 1, 2, 2, and 2. The RK4 update $\frac{1}{6}[d^{(1)}+2d^{(2)}+2d^{(3)}+d^{(4)}]$ achieves its fourth-order accuracy through the precise cancellation of relatively large errors (this cancellation works even when u is a vector and f is a nonlinear function). However, the necessity for these delicate cancellations to take place can lead to a problem if we have used the MOL concept to discretize a time-dependent equation and if its solution $u(x,t)$

is known analytically at a boundary. The exact boundary values at the time levels associated with the internal RK stages will lack the errors that the RK scheme expects. Through the spatial coupling, the critical error cancellations are then upset at locations near the boundary. Keeping a fixed grid in space and refining in time only, RK4 will (eventually) converge like $O(k^4)$ (this is how most MOL-ODE package runs are conducted - very convenient, but not optimally efficient). The boundary problem arises if we use (say) RK4 in time together with FD4 in space, and then refine the space and time steps together. The spatial coupling then increases as k decreases, and RK4 drops back to second order unless special attention is paid to the boundary implementation.

Table G.1-4 summarizes some different ways that a boundary condition $u(0,t)=g(t)$ might be implemented during the internal stages of the RK4 scheme (G.1-1). The subscript 0 on u signifies that we are discussing the zeroth (i.e. the boundary) component of the u vector that resulted from the spatial discretization. The last of the methods (involving derivatives of the boundary function $g(t)$) is described in Carpenter et al. (1993). It can heuristically be deduced as follows: In MOL, we first discretize in space and then apply an ODE solver (e.g. RK4) in time. Reversing this - that is, conceptually RK4-advancing the governing equation first - allows the space operator to be exchanged for a time derivative at the boundary (making use of the governing equation also at the boundary). This argument leads to the first formulation; the next two are immediate consequences. Pathria (1994) gives implementation details for a nonlinear extension of this approach.

G.2. Linear multi-step methods

Figure G.2-1 sketches a variety of possible stencils. An unknown value for f at the new time level $t+k$ implies that the method is implicit. The stencils extend back to previous time levels, for either u or f - but not deep back for both u and f. The reason stems from the following result.

First Dahlquist stability barrier. *The order of accuracy p of a stable s-step linear multi-step formula satisfies*

$$p \le \begin{cases} s+2 & \text{if } s \text{ is even,} \\ s+1 & \text{if } s \text{ is odd,} \\ s & \text{if the formula is explicit.} \end{cases}$$

A number of proofs are available. The idea of "order stars" (see Wanner, Hairer, and Nørsett 1978) leads to a particularly elegant one (Iserles and Nørsett 1984).

Table G.1-4. *Different strategies for implementing a Dirichlet boundary condition $u(0,t) = g(t)$ during the internal stages of an RK4-MOL time stepping process*

Description of boundary procedure	Boundary values u_0 to use at the four internal stages	Order of accuracy[a]	Comments
Use the analytic value in time that is associated with each internal stage.	$\begin{bmatrix} u_0^{(1)} \\ u_0^{(2)} \\ u_0^{(3)} \\ u_0^{(4)} \end{bmatrix} = \begin{bmatrix} g(t_n) \\ g(t_n+\frac{k}{2}) \\ g(t_n+\frac{k}{2}) \\ g(t_n+k) \end{bmatrix}$	2	Boundary values lack the 'errors' expected during the different stages – the mismatches ruin normal cancellation of errors to final 4th order of accuracy.
Implement $u(0,t)=g(t)$ as $u_t(0,t)=g'(t)$ and solve with RK4 together with remaining equations.	$\begin{bmatrix} u_0^{(1)} \\ u_0^{(2)} \\ u_0^{(3)} \\ u_0^{(4)} \end{bmatrix} = \begin{bmatrix} g(t_n) \\ g(t_n) \\ g(t_n) \\ g(t_n) \end{bmatrix} + \begin{bmatrix} 0 \\ \frac{k}{2}g'(t_n) \\ \frac{k}{2}g'(t_n+\frac{k}{2}) \\ kg'(t_n+\frac{k}{2}) \end{bmatrix}$	2	Like the case above, full (4th-order) accuracy if space grid is independent of k; however, loses accuracy if spatial grid is refined when k is decreased.
Obtain boundary stage values through spatial extrapolation.	$u_0^{(1)} = g(t_n)$ $u_0^{(2)}, u_0^{(3)}, u_0^{(4)}$ through high-order spatial extrapolation to the boundary from the interior of the domain	4 (p)	Full accuracy in all cases – but time-step stability condition might have become more restrictive.
Use governing equation (i.e. not boundary condition) at the boundary during internal stages.	$u_0^{(1)} = g(t_n)$ $u_0^{(2)}, u_0^{(3)}, u_0^{(4)}$ through RK advancement of one-sided approximation of the interior equations	4 (p)	

Use, for internal stages, updates based on differentiated versions of the boundary function.	$\begin{bmatrix} u_0^{(1)} \\ u_0^{(2)} \\ u_0^{(3)} \\ u_0^{(4)} \end{bmatrix} = \begin{bmatrix} g(t_n) \\ g(t_n) \\ g(t_n) \\ g(t_n) \end{bmatrix} + \begin{bmatrix} 0 \\ \frac{k}{2}(u_0^{(1)})' \\ \frac{k}{2}(u_0^{(2)})' \\ k(u_0^{(3)})' \end{bmatrix} = \begin{bmatrix} 1 & 0 & 0 & 0 \\ 1 & \frac{1}{2} & 0 & 0 \\ 1 & \frac{1}{2} & \frac{1}{4} & 0 \\ 1 & 1 & \frac{1}{2} & \frac{1}{4} \end{bmatrix} \begin{bmatrix} g(t_n) \\ kg'(t_n) \\ k^2 g''(t_n) \\ k^3 g'''(t_n) \end{bmatrix}$ Can be approximated in function values of g alone – for example: $\begin{bmatrix} u_0^{(1)} \\ u_0^{(2)} \\ u_0^{(3)} \\ u_0^{(4)} \end{bmatrix} = \begin{bmatrix} 0 & 1 & 0 & 0 \\ -\frac{1}{6} & \frac{3}{4} & \frac{1}{2} & -\frac{1}{12} \\ \frac{1}{12} & \frac{1}{4} & \frac{3}{4} & -\frac{1}{12} \\ -\frac{1}{12} & \frac{1}{4} & \frac{3}{4} & \frac{1}{12} \end{bmatrix} \begin{bmatrix} g(t_n - k) \\ g(t_n) \\ g(t_n + k) \\ g(t_n + 2k) \end{bmatrix}$	4 (p)	As formulated here, accurate only for linear problems with time-independent coefficients. Method can be generalized to time-dependent and nonlinear functions. The recommended approach.

Note: The symbol $u_0^{(i)}$ denotes the boundary values to use when calculating $d^{(i)}$, $i = 1, 2, 3, 4$, in equation (G.1-1).

[a] "4 (p)" indicates that if the RK scheme is enhanced from order 4 to order p, then the boundary procedure would feature the same gain in accuracy.

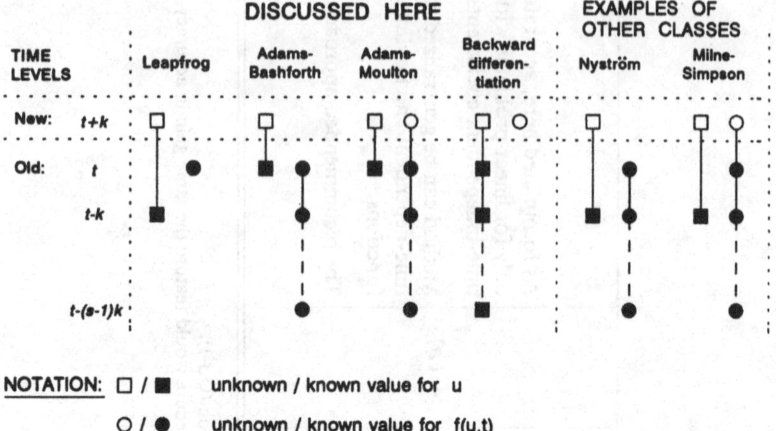

Figure G.2-1. Stencils of some different types of linear multi-step methods for solving $u' = f(u, t)$.

This result tells us that it is futile to try to reach high orders by letting both u and f extend deep back. Such schemes are unconditionally unstable.

a. Leapfrog

The leapfrog (LF) scheme is often used because of its simplicity. This is a slightly dangerous choice, since its stability domain covers no area – it stretches between $+i$ and $-i$ on the imaginary axis. This rules out LF for all problems but purely hyperbolic ones, approximated by antisymmetric FD stencils (such as the Fourier–PS method). However, in such cases it can be a convenient scheme, entirely free of dissipative (but not dispersive) errors. Among the demonstration problems in this book, the results in Figures 4.2-2, 7.1-2(b), 8.1-1, 8.2-1, 8.2-3, and 8.4-2 were obtained using LF time stepping.

The leapfrog time step for an FD approximation to a hyperbolic PDE, say, $u_t = a(x)u$, takes the form

$$\frac{u(x,t+k)-u(x,t-k)}{2k} = a(x)\times\begin{bmatrix}\text{approximation to } u_x \\ \text{calculated at } (x,t)\end{bmatrix}.$$

The left-hand side can be seen as the average of a forward and a backward approximation:

$$\frac{1}{2}\left\{\frac{u(x,t+k)-u(x,t)}{k}+\frac{u(x,t)-u(x,t-k)}{k}\right\} = \cdots. \qquad \text{(G.1-2)}$$

Iserles (1986) notes that when $a(x)$ is of fixed sign, say $a(x) > 0$ (for systems whose characteristics all go either to the right or to the left – a typical situation locally for quantities convected with a fluid flow), accuracy can be gained by making the two approximations in (G.1-2) at adjacent rather than at identical x positions:

$$\frac{1}{2}\left\{\frac{u(x+h,t+k)-u(x+h,t)}{k} + \frac{u(x,t)-u(x,t-k)}{k}\right\} = a(x)\times\left[\begin{array}{c}\text{approximation to } u_x \\ \text{centered at } (x+h/2,t)\end{array}\right].$$

All schemes of this kind are dissipation-free (since they are time-reversible). Roe (1994) extends the idea to schemes in several space dimensions.

b. Adams-Bashforth/Adams-Moulton

Some AB and AM formulas of increasing order are shown in Table G.2-1, where ∇ denotes the backward different operator

$$\nabla f_n = f_n - f_{n-1},\ \nabla^2 f_n = f_n - 2f_{n-1} + f_{n-2},\ \dots .$$

The coefficients

$$\gamma_{-1}, \gamma_0, \gamma_1, \gamma_2, \dots = 0, 1, \tfrac{1}{2}, \tfrac{5}{12}, \tfrac{3}{8}, \tfrac{251}{720}, \tfrac{95}{288}, \dots$$

appear as the top row in the chart shown as Table G.2-2. The values in the first column are $1/i$, $i = 1, 2, \dots$, and the values in subsequent columns $j = 1, 2, 3, \dots$ are obtained through the recursion

$$c_{i,j} = c_{i,j-1} - c_{i+1,j-1}/j$$

(which can be generalized to variable grid spacings; cf. Shampine and Gordon, 1975). The very lowest-order AB and AM methods are known also as:

- AB1 - forward Euler;
- AM1 - backward Euler;

Table G.2-1. *Explicit expressions for low-order AB/AM schemes*

Order	AB (predictor)	AM (corrector)
1	$u_{n+1} = u_n + kf_n$	
2	$u_{n+1} = u_n + \frac{k}{2}[3f_n - f_{n-1}]$	$u_{n+1} = u_n + \frac{k}{2}[f_{n+1} + f_n]$
3	$u_{n+1} = u_n + \frac{k}{12}[23f_n - 16f_{n-1} + 5f_{n-2}]$	$u_{n+1} = u_n + \frac{k}{12}[5f_{n+1} + 8f_n - f_{n-1}]$
4	$\vdots$	$u_{n+1} = u_n + \frac{k}{24}[9f_{n+1} + 19f_n - 5f_{n-1} + f_{n-2}]$
$\vdots$	$\vdots$	$\vdots$
p	$u_{n+1} = u_n + k\sum_{i=0}^{p-2} \gamma_i \nabla^i f_n$	$u_{n+1} = u_n + k\sum_{i=0}^{p-1} (\gamma_i - \gamma_{i-1})\nabla^i f_{n+1}$

Note: The bottom line shows the general forms in cases of arbitrary orders.

Table G.2-2. *Chart for generating the coefficients in AB and AM formulas*

i	j = 0	1	2	3	...
1	1	$\frac{1}{2}$	$\frac{5}{12}$	$\frac{3}{8}$	...
2	$\frac{1}{2}$	$\frac{1}{6}$	$\frac{1}{8}$	⋮	
3	$\frac{1}{3}$	$\frac{1}{12}$	⋮		
4	$\frac{1}{4}$	⋮			
⋮	⋮				

- AM2 – Crank–Nicolson, implicit midpoint rule, and the $s=1$, $p=2$ implicit RK method.

The AB methods are explicit. The AM methods are implicit, but they have, for the same order of accuracy, smaller error coefficients and larger stability domains. One (of several) possible ways to combine the best of both methods into a single procedure of order p is the following:

(1) use AB of order $p-1$ to get a predicted value u_{n+1}^*;
(2) use this predicted value u_{n+1}^* to calculate a predicted value f_{n+1}^*;
(3) use f_{n+1}^* in the right-hand side of AM of order p, obtaining a corrected value u_{n+1}; and
(4) calculate a corrected value f_{n+1} (for use in subsequent time steps).

This predictor–corrector procedure gives for different p the stability domains shown in Figure G.2-2. These domains are smaller than those for explicit RK methods, and they decrease as p is increased. On the other hand, only two function evaluations are needed per time step, compared to s ($\geq p$) for explicit RK methods.

In order to simplify error control, AB/AM packages often use predictors and correctors of the same order of accuracy. The choice of orders, step sizes, etc. are handled automatically within ODE packages, and are of little direct concern to MOL users.

Predictor–corrector schemes need not be AB/AM variations. For example, Hyman (1979) notes that the leapfrog predictor $u_{n+1}^* = u_{n-1}+2kf_n$, followed by the corrector

$$u_{n+1} = \frac{4}{5}u_n + \frac{1}{5}u_{n-1} + \frac{2k}{5}(f_{n+1}^* + 2f_n),$$

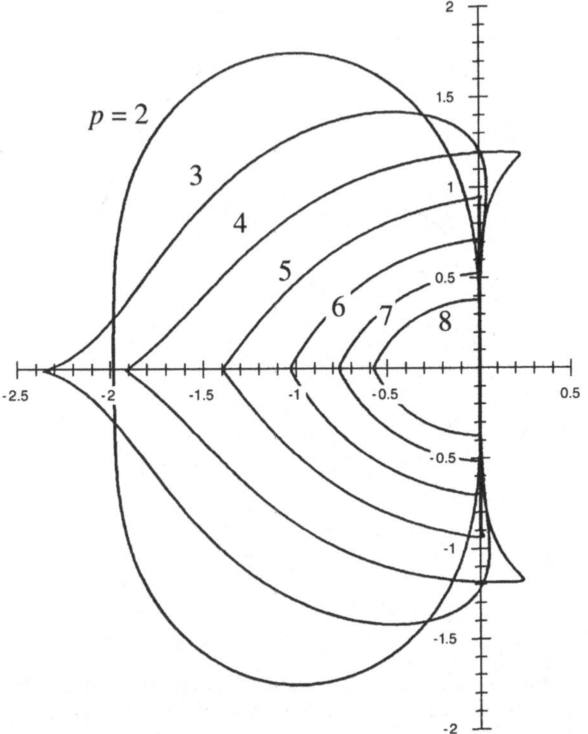

Figure G.2-2. Stability domains for an Adams–Bashforth/Adams–Moulton predictor–corrector scheme, shown for orders $2 \le p \le 8$ (predictor of order $p-1$, corrector of order p).

achieves third-order accuracy and has a stability domain extending along the imaginary and negative real axes to $\pm\frac{3}{2}i$ and $-\frac{3}{2}$, respectively.

c. *Backward differentiation methods*

With the notation used in Table 3.1-3 and numerical values from Table 3.1-2, the backward differentiation formulas (BDFs) take the form

$$c^1_{p,0}u_{n+1} + c^1_{p,1}u_n + c^1_{p,2}u_{n-1} + \cdots + c^1_{p,p}u_{n-(p-1)} = f_{n+1}.$$

The stability regions for $p = 1, 2, \ldots, 6$ are shown in Figure G.2-3, with the area near the origin enlarged in Figure G.2-4. In contrast to the previous stability diagrams, the BDF methods are stable *outside* the closed curves shown. Up to sixth order, the complete negative real axis falls within the stable domains. For $p \ge 7$, the BDF schemes become unconditionally unstable – no neighborhood of the origin is covered.

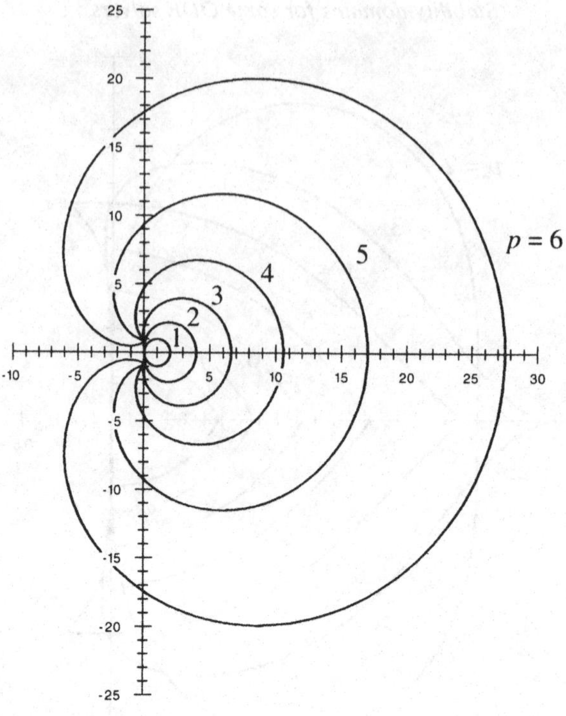

Figure G.2-3. Stability diagram for backward differentiation methods, $p \leq 6$. Note: the methods are stable *outside* the closed regions.

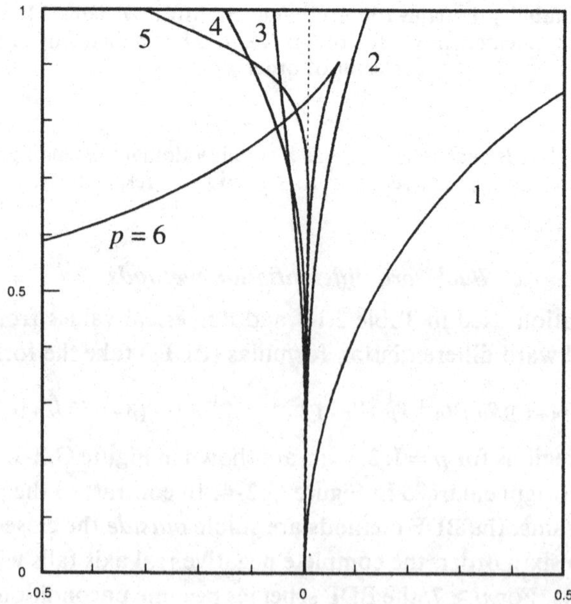

Figure G.2-4. Detail near the origin of the stability regions shown in Figure G.2-3.

Table G.3-1. *Stability of some ODE solvers when all EVs are on the imaginary axis*

	Conditionally stable	Unstable
Explicit RK	$p=s=3,4$, only a few with $p>4$	$p=s=1,2$, most with $p>4$
AB	$p=3,4,\ 7,8,\ 11,12,\ \dots$	$p=1,2,\ 5,6,\ 9,10,\ \dots$
AM	$p=1,2,\ 5,6,\ 9,10,\ \dots$	$p=3,4,\ 7,8,\ 11,12,\ \dots$
AB/AM[a]	most	some
BDF[b]	$p=1,2,\ 5,6$	$p=3,4$, all with $p>6$

[a] Predictor/corrector.
[b] For this solver, $p=1,2$ is unconditionally stable.

G.3. Extent of stability domains along the imaginary axis

In cases of FD or PS approximations of periodic, purely convective problems, the eigenvalues of the DMs that arise in MOL discretizations fall precisely along the imaginary axis. This is a situation that most ODE packages are not particularly designed for. Only some of the common ODE techniques work satisfactorily in this case. Table G.3-1 summarizes this situation schematically.

H

Energy estimates

Different variations of the "energy method" can be used to show that PDEs are well posed, to show that discrete approximations are stable, and to establish (global) convergence rates under mesh refinements. The energy approach is very broadly applicable, and can handle many cases that include: boundary conditions; variable coefficients (and nonlinearities); and nonperiodic PS methods.

However, this flexibility and power comes at a price of often significant technical difficulty. This appendix is intended to provide only a first flavor of this rich subject to readers who are unfamiliar with it. For this purpose, we here consider five examples, all relating to the heat equation on the interval $[-1,1]$. More systematic descriptions can be found in Richtmyer and Morton (1967), Gustafsson et al. (1995), and (for PS methods) Canuto et al. (1988).

Example 1. Show that the heat equation

$$u_t = u_{xx}$$

$$\text{subject to} \qquad \text{initial condition: } u(x,0) = f(x), \qquad \text{(H-1)}$$

$$\text{boundary conditions: } u(-1,t) = u(1,t) = 0$$

is *well posed* – that is, some norm of the solution can be bounded by the initial data:

$$\|u(\cdot,t)\| \le c_1 e^{c_2 t} \|f(\cdot)\|, \quad c_1, c_2 > 0.$$

Choosing to consider the L^2 norm $\|u\|^2 = \int_{-1}^{1} u^2\,dx$, we obtain

$$\frac{\partial}{\partial t}\|u\|^2 = 2\int_{-1}^{1} u_t u\,dx \qquad \text{substitute } u_t = u_{xx}$$

$$= 2\int_{-1}^{1} u_{xx} u\,dx \qquad \text{partial integration; end contributions vanish}$$

$$= -2\int_{-1}^{1} (u_x)^2\, dx \le 0.$$

In other words,

$$\|u(\cdot,t)\|^2 \le \|f(\cdot)\|^2.$$

Example 2. Show well-posedness when (H-1) is generalized to

$$u_t = a(x,t)u_{xx} + b(x,t)u_x + c(x,t),$$

where $a(x,t)$ is differentiable and satisfies $a(x,t) \ge a_0 > 0$.

Using the notation

$$(u,v) = \int_{-1}^{1} \bar{u}\cdot v\, dx, \qquad \|u\|^2 = (u,u),$$

complex conjugation of first argument is often appropriate in cases of u and v complex; this is not an essential issue in the present context

we obtain (following Gustafsson et al. 1995):

$$\frac{\partial}{\partial t}\|u\|^2 = I + II + III,$$

where

$$I = (u, au_{xx}) + (au_{xx}, u)$$
partial integration

$$= -(u_x, au_x) - (u, a_x u_x) - (au_x, u_x)$$
use $|(u,av)| \le \|a\|_\infty \|u\|\|v\|$, where $\|a\|_\infty = \sup_x |a|$ (a generalization of the Cauchy-Schwarz inequality $|(u,v)| \le \|u\|\|v\|$)

$$-(a_x u_x, u) + a(\bar{u}u_x + \bar{u}_x u)|_{-1}^{1}$$
boundary terms vanish

$$\le -2(u_x, au_x) + 2\|a_x\|_\infty \|u\|\|u_x\|$$

$$\le -2a_0\|u_x\|^2 + 2\|a_x\|_\infty \sqrt{2/a_0}\,\|u\|\sqrt{a_0/2}\,\|x\|$$
use the inequality $2ab \le a^2 + b^2$

$$\le -\tfrac{3}{2}a_0\|u_x\|^2 + 2(\|a_x\|_\infty^2/a_0)\|u\|^2;$$

$$II = (u, bu_x) + (bu_x, u)$$
use same inequalities as before

$$\le 2\|b\|_\infty \|u\|\|u_x\|$$

$$\le (\|b\|_\infty^2/a_0)\|u\|^2 + a_0\|u_x\|^2;$$

$$III = (u, cu) + (cu, u) \le 2\|c\|_\infty \|u\|^2.$$

Thus

$$\frac{\partial}{\partial t}\|u\|^2 \le -\frac{1}{2}a_0\|u_x\|^2 + \alpha\|u\|^2$$
where $\alpha = \dfrac{2\|a_x\|_\infty^2 + \|b\|_\infty^2}{a_0} + 2\|c\|_\infty$

$$\le \alpha\|u\|^2,$$

and we obtain directly, or by referring to Gronwall's lemma

$$\phi'(t) \le \alpha\phi(t) + g(t) \Rightarrow \phi(t) \le e^{\alpha t}\phi(0) + \int_0^t g(s)e^{\alpha(t-s)}\,ds$$

in its special case of $g(t) \equiv 0$,

$$\|u(\cdot,t)\|^2 \le e^{\alpha t}\|f(\cdot)\|^2.$$

Example 3. Show that the forward Euler–FD2 scheme

$$\frac{u(x,t+k)-u(x,t)}{k} = \frac{u(x-h,t)-2u(x,t)+u(x+h,t)}{h^2}, \qquad \text{(H-2)}$$

when applied to (H-1), is numerically stable for some values of $\lambda = k/h^2$ (cf. Example 3 in Section 4.5).

Max-norm stability: Few FD schemes do not allow any growth in the max-norm. In such cases, stability can often be proven very easily. We write the scheme (H-2) as

$$u(x,t+k) = \lambda u(x-h,t) + (1-2\lambda)u(x,t) + \lambda u(x+h,t).$$

If $0 \le \lambda \le \frac{1}{2}$, then the coefficients λ, $(1-2\lambda)$, and λ are all nonnegative and add up to 1. Therefore

$$|u(x,t+k)| \le \max\{|u(x-h,t)|, |u(x,t)|, |u(x+h,t)|\}$$

and

$$\|u(\cdot,t)\|_\infty \le \|f(\cdot)\|_\infty \quad (\text{for } \lambda = k/h^2 \le \tfrac{1}{2}).$$

L^2-norm stability: The general procedure we employ here is quite typical for many FD schemes. We first introduce some convenient notation (somewhat tailored to $[-1,1]$; $h = 2/N$):

$$t_m = mk, \quad x_i = -1 + ih, \quad u_i^m = u(x_i, t_m), \quad \|u^m\|_{p,s}^2 = h\sum_{i=p}^{s}(u_i^m)^2;$$

$$D_+u_i = \frac{1}{h}[u_{i+1} - u_i], \quad \text{that is,} \quad \|D_+u\|_{0,N-1}^2 = \frac{1}{h}\sum_{i=0}^{N-1}(u_{i+1}-u_i)^2.$$

The FD scheme (H-2) can now be written

$$u_i^{m+1} - u_i^m = \frac{k}{h^2}(u_{i-1}^m - 2u_i^m + u_{i+1}^m). \qquad \text{(H-3)}$$

We multiply this by $h(u_i^{m+1} + u_i^m)$ and sum over i to obtain

$$\|u^{m+1}\|_{0,N}^2-\|u^m\|_{0,N}^2$$

$$=\frac{k}{h}\sum_{i=1}^{N-1}(u_i^{m+1}+u_i^m)(u_{i-1}^m-2u_i^m+u_{i+1}^m)$$

use partial summation; with $u_0=v_0=u_N=v_N=0$,

$$\sum_{i=1}^{N-1}v_i(u_{i-1}-2u_i+u_{i+1}) = -\sum_{i=0}^{N-1}(v_{i+1}-v_i)(u_{i+1}-u_i)$$

$$=-\frac{k}{h}\left[\sum_{i=0}^{N-1}(u_{i+1}^{m+1}-u_i^{m+1})(u_{i+1}^m-u_i^m)\right]$$

use the inequality $ab\le(a^2+b^2)/2$

$$-\frac{k}{h}\left[\sum_{i=0}^{N-1}(u_{i+1}^m-u_i^m)^2\right]$$

$$\le\frac{k}{2h}\left[\sum_{i=0}^{N-1}(u_{i+1}^{m+1}-u_i^{m+1})^2+\sum_{i=0}^{N-1}(u_{i+1}^m-u_i^m)^2\right]$$

$$-k\|D_+u^m\|_{0,N-1}^2$$

$$=\frac{k}{2}(\|D_+u^{m+1}\|_{0,N-1}^2-\|D_+u^m\|_{0,N-1}^2).$$

We now introduce the notation $S^m=\|u^m\|_{0,N}^2-(k/2)\|D_+u^m\|_{0,N-1}^2$. Because of the preceding inequality, this quantity satisfies

$$S^{m+1}-S^m=\|u^{m+1}\|_{0,N}^2-\|u^m\|_{0,N}^2-\frac{k}{2}(\|D_+u^{m+1}\|_{0,N-1}^2-\|D_+u^m\|_{0,N-1}^2)$$
$$\le 0;$$

that is, it cannot grow with m. Stability of the numerical scheme (H-2) will follow if we can show that this bound on S^m implies a bound on $\|u^m\|_{0,N}^2$ (note that no condition on $\lambda=k/h^2$ has entered yet - it must come in now). We have

$$S^m=\|u^m\|_{0,N}^2-\frac{\lambda h}{2}\sum_{i=0}^{N-1}(u_{i+1}^m-u_i^m)^2$$

use $(a-b)^2=a^2-2ab+b^2 \le 2(a^2+b^2)$

$$\ge\|u^m\|_{0,N}^2-2\lambda h\sum_{i=0}^{N}(u_i^m)^2$$

assume now that $\lambda\le\frac{1}{2}(1-\epsilon)$ for some $\epsilon>0$

$$=\|u^m\|_{0,N}^2-(1-\epsilon)\|u^m\|_{0,N}^2=\epsilon\|u^m\|_{0,N}^2.$$

Therefore,

$$\|u^m\|_{0,N}^2\le\frac{1}{\epsilon}S^m\le\frac{1}{\epsilon}S^0\le\frac{1}{\epsilon}\|f\|_{0,N}^2$$

holds if $\lambda\le\frac{1}{2}(1-\epsilon)$. Note that this argument gave a sufficient, but not necessary, condition for stability - it failed to establish stability when $\lambda=\frac{1}{2}$.

The case $\lambda = \frac{1}{2}$ can be handled separately. Equation (H-3) in this case becomes

$$u_i^{m+1} = \tfrac{1}{2}(u_{i-1}^m + u_{i+1}^m).$$

Squaring both sides yields

$$
\begin{aligned}
(u_i^{m+1})^2 &= \tfrac{1}{4}(u_{i-1}^m + u_{i+1}^m)^2 && \text{use the inequality } (a+b)^2 \le 2(a^2+b^2) \\
&\le \tfrac{1}{2}((u_{i-1}^m)^2 + (u_{i+1}^m)^2); && \text{sum over } i \\
\|u^{m+1}\|_{0,N}^2 &\le \tfrac{1}{2}\|u^m\|_{0,N}^2 + \tfrac{1}{2}\|u^m\|_{0,N}^2 \\
&= \|u^m\|_{0,N}^2.
\end{aligned}
$$

Example 4. Show that equation (H-1) remains well posed when the u_{xx} term is replaced by a Legendre- or Chebyshev-PS approximation.

At each time instant t, $u(x,t)$ is approximated by an Nth-degree polynomial $u_N(x,t)$ in x, with coefficients depending on t. Let x_i and w_i be (resp.) the knots and weights for the Legendre–Gauss–Lobatto formula (as shown in Table 4.7-1). The PS collocation procedure implies that

$$\frac{\partial u_N(x_i,t)}{\partial t} = \frac{\partial^2 u_N(x_i,t)}{\partial x^2}$$

holds at each of the quadrature knots x_i, $i = 0, 1, \ldots, N$. Therefore

$$
\begin{aligned}
&\frac{\partial}{\partial t}\sum_{i=0}^{N}[u_N(x_i,t)]^2 w_i \\
&= 2\sum_{i=0}^{N}\frac{\partial u_N(x_i,t)}{\partial t}u_N(x_i,t)w_i && \text{replace } t \text{ derivatives with } x \text{ derivatives} \\
&= 2\sum_{i=0}^{N}\frac{\partial^2 u_N(x_i,t)}{\partial x^2}u_N(x_i,t)w_i && \Sigma = \textstyle\int \text{ since } (\partial^2 u_N/\partial x^2)u_N \text{ is a polynomial of degree } 2N-2 \text{ (the Gaussian quadrature formula is exact for all polynomials up to and including degree } 2N-1) \\
&= 2\int_{-1}^{1}\frac{\partial^2 u_N(x,t)}{\partial x^2}u_N(x,t)\,dx && \text{partial integration} \\
&= -2\int_{-1}^{1}\left(\frac{\partial u_N(x,t)}{\partial x}\right)^2 dx \le 0.
\end{aligned}
$$

In the Chebyshev case, we use instead the Gauss–Chebyshev–Lobatto nodes and weights. At the partial integration step, a complication arises since a weight function $w(x) = 1/\sqrt{1-x^2}$ is now also present:

$$
\begin{aligned}
&\vdots \\
&= 2\int_{-1}^{1}\frac{\partial^2 u_N(x,t)}{\partial x^2}u_N(x,t)w(x)\,dx && \text{the proof for the first inequality is relatively tricky – see e.g. Canuto et al. (1988, Sec. 11.1.2) for details}
\end{aligned}
$$

$$\leq -\frac{1}{2}\int_{-1}^{1}\left(\frac{\partial u_N(x,t)}{\partial x}\right)^2 w(x)\,dx \leq 0.$$

That the nodes were chosen according to Gaussian quadrature formulas was critical for the success of the particular argument used in this example. As noted in Section 4.7, these node choices seem to be much less critical as far as actual PS performance is concerned.

Example 5. Show that the Chebyshev–PS approximation to the heat equation (H-1) converges exponentially fast (for increasing number of node points N).

In order to estimate the error, we need to consider two different approximations to the analytic solution $u(x,t)$ of equation (H-1):

$\tilde{u}_N(x,t)$, the Nth-degree polynomial in x that interpolates $u(x,t)$ at the Chebyshev node points x_i, $i=0,1,\ldots,N$; and

$u_N(x,t)$, the Chebyshev–PS solution to (H-1), which is also a polynomial of degree N in x.

The demonstration that $u_N(x,t)$ is exponentially close to $u(x,t)$ is carried out in two steps.

1. *Show that the error $u-u_N$ at the nodes x_i can be bounded in terms of* $r=(\partial^2/\partial x^2)(u-\tilde{u}_N)$. At the nodes x_i, $i=0,1,\ldots,N$,

$$\frac{\partial \tilde{u}_N}{\partial t}-\frac{\partial^2 \tilde{u}_N}{\partial x^2}=r, \qquad \text{since } \frac{\partial \tilde{u}_N}{\partial t}=\frac{\partial u}{\partial t}=\frac{\partial^2 u}{\partial x^2}$$

$$\frac{\partial u_N}{\partial t}-\frac{\partial^2 u_N}{\partial x^2}=0. \qquad \text{Chebyshev–PS method}$$

The difference $e_N=\tilde{u}_N-u_N$ is an Nth-degree polynomial that satisfies $e_N(x,0)\equiv 0$. Subtraction of the two equations just displayed gives (at x_i, $i=0,1,\ldots,N$)

$$\frac{\partial e_N}{\partial t}-\frac{\partial^2 e_N}{\partial x^2}=r;$$

that is (with w_i denoting the Chebyshev–GQ weights),

$$\frac{\partial}{\partial t}\sum_{i=0}^{N}[e_N(x_i,t)]^2 w_i$$

$$=2\sum_{i=0}^{N}\frac{\partial e_N(x_i,t)}{\partial t}e_N(x_i,t)w_i$$

$$=2\sum_{i=0}^{N}\frac{\partial^2 e_N}{\partial x^2}(x_i,t)e_N(x_i,t)w_i+2\sum_{i=0}^{N}r(x_i,t)e_N(x_i,t)w_i$$

swap the first Σ to $\int$, then note (as in Example 4) that $\int \le 0$; in the second Σ, use $2ab \le a^2 + b^2$

$$\le \sum_{i=0}^{N} [r(x_i,t)]^2 w_i + \sum_{i=0}^{N} [e_N(x_i,t)]^2 w_i.$$

Gronwall's lemma (see Example 2) gives

$$\sum_{i=0}^{N} [e_N(x_i,t)]^2 w_i \le e^t \int_0^t \sum_{i=0}^{N} [r(x_i,s)]^2 w_i \, ds.$$

At the nodes (where $e_N = u - u_N$), e_N has been bounded in terms of r.

2. *Estimate the size of* $r = (\partial^2/\partial x^2)(u - \tilde{u}_N)$ *at the nodes* x_i, $i = 0, 1, \ldots,$ N. This estimate can be carried out in a formal manner, as for example in Canuto et al. (1988, Sec. 10.1.2), giving specific constants and decay rates that are dependent on the smoothness of $u(x,t)$. For a more heuristic argument, we note that r, at the nodes, expresses the error when the second derivative of $u(x,t)$ is approximated in the standard PS manner. The accuracy of this procedure was the topic of Sections 3.4 and 4.1. Because $u(x,t)$ is analytic for $t > 0$, r will decay exponentially fast with N.

References

S. Abarbanel, D. Gottlieb, and E. Tadmor (1986), "Spectral methods for discontinuous problems", in *Numerical Methods for Fluid Dynamics II* (K. W. Morton and M. J. Baines, eds.), Clarendon Press, Oxford, pp. 129–53.

M. J. Ablowitz, B. M. Herbst, and C. M. Schober (1995a). "Numerical simulation of quasi-periodic solutions of the sine-Gordon equation", *Physica D* (to appear).

M. J. Ablowitz, B. M. Herbst, and C. M. Schober (1995b), "On the numerical solution of the sine-Gordon equation. I. Integrable discretizations and homoclinic orbits", *J. Comput. Phys.* (submitted).

M. J. Ablowitz, B. M. Herbst, and C. M. Schober (1995c), "On the numerical solution of the sine-Gordon equation. II. Numerical methods" (in preparation).

M. J. Ablowitz and J. F. Ladik (1976), "A nonlinear difference scheme and inverse scattering", *Stud. Appl. Math.* 55: 213–29.

M. J. Ablowitz and H. Segur (1981), *Solitons and the Inverse Scattering Transform,* SIAM, Philadelphia.

R. C. Agarwal and S. Burrus (1975), "Number theoretic transforms to implement fast digital convolution", *Proc. IEEE* 63: 550–60.

R. M. Alford, K. R. Kelly, and D. M. Boore (1974), "Accuracy of finite-difference modeling of the acoustic wave equation", *Geophysics* 39: 834–42.

B. Alpert and V. Rokhlin (1991), "A fast algorithm for the evaluation of Legendre expansions", *SIAM J. Sci. Stat. Comp.* 12: 158–79.

A. Aoyagi (1995), "Nonlinear leapfrog instability for Fornberg's pattern", *J. Comput. Phys.* (submitted).

A. Aoyagi and K. Abe (1989), "Parametric excitation of computational modes inherent to leap-frog schemes", *J. Comput. Phys.* 83: 447–62.

A. Arakawa (1966), "Computational design for long-term numerical integration of the equations of fluid motion: Two dimensional incompressible flow", *J. Comput. Phys.* 1: 119–43.

D. H. Bailey and P. N. Swarztrauber (1991), "The fractional Fourier transform and applications", *SIAM Review* 33: 389–404.

D. H. Bailey and P. N. Swarztrauber (1994), "A fast method for the numerical evaluation of continuous Fourier and Laplace transforms", *SIAM J. Sci. Comput.* 15: 1105–10.

D. Barton, I. M. Willers, and R. V. M. Zahar (1971), "An implementation of the Taylor series method for ordinary differential equations", *Comput. J.* 14: 243–8.

G. K. Batchelor (1953), *Theory of Homogeneous Turbulence,* Cambridge University Press.

A. Bayliss, D. Gottlieb, B. J. Matkowsky, and M. Minkoff (1989), "An adaptive pseudospectral method for reaction diffusion problems", *J. Comput. Phys.* 81: 421–43.

A. Bayliss and E. Turkel (1992), "Mappings and accuracy for Chebyshev pseudospectral approximations", *J. Comput. Phys.* 101: 349–59.

G. E. Bell (1986), "Richardson's method - a rediscovery", *Bull. Inst. Math. Appl.* 22: 41–3.

S. L. Belousov (1962), *Tables of Normalized Associated Legendre Polynomials,* MacMillan, New York.

C. Bernardi and Y. Maday (1991), "Some spectral approximations of one-dimensional fourth-order problems", in *Progress in Approximation Theory* (P. Nevai and A. Pinkus, eds.), Academic Press, San Francisco, pp. 43–116.

G. Beylkin (1992), "On the representation of operators in bases of compactly supported wavelets", *SIAM J. Numer. Anal.* 6: 1716–40.

G. Beylkin and M. E. Brewster (1995), "Fast numerical algorithms using wavelet basis on an interval" (in preparation).

G. Beylkin, R. Coifman, and V. Rokhlin (1991), "Fast wavelet transforms and numerical algorithms I", *Comm. Pure Appl. Math.* 44: 141–83.

M. Bhattacharya and R. C. Agarwal (1984), "Number theoretic techniques for computation of digital convolution", *IEEE Trans. Acoust. Speech and Signal Proc.* 32: 507–11.

L. I. Bluestein (1970), "A linear filtering approach to the computation of the discrete Fourier transform", *IEEE Trans. Audio Electroacoust.* 18: 451–5.

J. P. Boyd (1987), "Spectral methods using rational basis functions on an infinite interval", *J. Comput. Phys.* 69: 112–42.

J. P. Boyd (1989), *Chebyshev and Fourier Spectral Methods,* Springer-Verlag, New York.

J. P. Boyd (1992), "Multipole expansions and pseudospectral cardinal functions: a new generalization of the fast Fourier transform", *J. Comput. Phys.* 103: 184–6.

J. P. Boyd (1994), "The rate of convergence of Fourier coefficients for entire functions of infinite order with application to the Weideman–Cloot sinh-mapping for pseudospectral computations on an infinite interval", *J. Comput. Phys.* 110: 360–72.

M. E. Brachet, D. I. Meiron, S. A. Orszag, B. G. Nickel, R. H. Morf, and U. Frisch (1983), "Small-scale structure of the Taylor–Green vortex", *J. Fluid Mech.* 130: 411–52.

K. S. Breuer and R. M. Everson (1992), "On the errors incurred calculating derivatives using Chebyshev polynomials", *J. Comput. Phys.* 99: 56–67.

G. L. Browning, J. J. Hack, and P. N. Swarztrauber (1988), "A comparison of three numerical methods for solving differential equations on the sphere", *Monthly Weather Review* 117: 1058–75.

G. L. Browning and H.-O. Kreiss (1989), "Comparison of numerical methods for the calculation of two-dimensional turbulence", *Math. Comput.* 52: 369–88.

H.-P. Bunge and J. R. Baumgardner (1995), "Mantle convection modeling on parallel virtual machines", *Computers in Physics* 9: 207–15.

W. Cai, D. Gottlieb, and C.-W. Shu (1989), "Essentially nonoscillatory spectral Fourier methods for shock wave calculation", *Math. Comput.* 52: 389–410.

W. Cai, D. Gottlieb, and C.-W. Shu (1992), "One-sided filters for spectral Fourier approximation of discontinuous functions", *SIAM J. Numer. Anal.* 29: 905–16.

C. Canuto, M. Y. Hussaini, A. Quarteroni, and T. Zang (1988), *Spectral Methods in Fluid Dynamics,* Springer-Verlag, New York.

C. Canuto and P. Pietra (1987), "Boundary and interface conditions with a FE preconditioner for spectral methods", Report no. 553, I.A.N., Pavia University, Italy.

C. Canuto and A. Quarteroni (1985), "Preconditioned minimal residual methods for Chebyshev spectral calculations", *J. Comput. Phys.* 60: 315–37.

C. Canuto and A. Quarteroni (1987), "On the boundary treatment in spectral methods for hyperbolic systems", *J. Comput. Phys.* 71: 100–10.

M. H. Carpenter, D. Gottlieb, S. Abarbanel, and W. S. Don (1993), "The theoretical accuracy of Runge–Kutta time discretizations for the initial boundary value problem: a careful study of the boundary error", ICASE Report no. 93-83, NASA Langley Research Center, Hampton, VA.

C. Cerjan, D. Kosloff, R. Kosloff, and M. Reshef (1985), "A nonreflecting boundary condition for discrete acoustic and elastic wave equations", *Geophysics* 50: 705–8.

T. F. Chan, R. Glowinski, J. Periaux, and O. B. Widlund, eds. (1989), *Domain Decomposition Methods,* SIAM, Philadelphia.

T. F. Chan, R. Glowinski, J. Periaux, and O. B. Widlund, eds. (1990), *Domain Decomposition Methods for Partial Differential Equations,* SIAM, Philadelphia.

T. F. Chan and T. Kerkhoven (1985), "Fourier methods with extended stability intervals for the Korteweg–de Vries equation", *SIAM J. Numer. Anal.* 22: 441–54.

T. F. Chan and T. P. Mathew (1994), "Domain decomposition algorithms", in *Acta Numerica 1994* (A. Iserles, ed.), Cambridge University Press, pp. 61–143.

S. Chen, G. D. Doolen, R. H. Kraichnan, and Z.-S. She (1993), "On statistical correlation between velocity increments and locally averaged dissipation in homogeneous turbulence", *Phys. Fluids A* 5: 458–63.

E. W. Cheney (1966), *Introduction to Approximation Theory,* McGraw-Hill, New York.

R. Clayton and B. Engquist (1977), "Absorbing boundary conditions for acoustic and elastic wave equations", *Bull. Seism. Soc. Amer.* 67: 1529–40.

A. Cloot and J. A. C. Weideman (1992), "An adaptive algorithm for spectral computations on unbounded domains", *J. Comput. Phys.* 102: 398–406.

L. Collatz (1960), *The Numerical Treatment of Differential Equations,* Springer-Verlag, Berlin.

J. W. Cooley and J. W. Tukey (1965), "An algorithm for the machine calculation of complex Fourier series", *Math. Comput.* 19: 297–301.

C. F. Corliss and Y. F. Chang (1982), "Solving ordinary differential equations using Taylor series", *ACM Trans. Math. Software* 8: 114–44.

R. Courant, K. Friedrichs, and H. Lewy (1928), "On the partial differential equations of mathematical physics" (German), *Math. Ann.* 100: 32–74; English translation (1967), *IBM Journal* 11: 215–34.

E. A. Coutsias, T. Hagstrom, and D. Torres (1994), "An efficient spectral method for ordinary differential equations with rational function coefficients", *Math. Comput.* (submitted).

J. Crank and P. Nicolson (1947), "A practical method for numerical integration of solutions of partial differential equations of heat-conduction type", *Proc. Cambridge Philos. Soc.* 43: 50–67.

G. Dahlquist (1956), "Convergence and stability in the numerical integration of ordinary differential equations", *Math. Scand.* 4: 33–53.

G. Dahlquist (1985), "33 years of numerical instability, part I", *BIT* 25: 188–204.

P. J. Davis (1975), *Interpolation and Approximation,* Dover, New York.

M. Deville and E. Mund (1985), "Chebyshev PS solution of second-order elliptic equations with finite element preconditioning", *J. Comput. Phys.* 60: 517–53.

G. A. Dilts (1985), "Computation of spherical harmonic expansion coefficients via FFT's", *J. Comput. Phys.* 57: 439–53.

W. S. Don and A. Solomonoff (1995), "Accuracy enhancement for higher derivatives using Chebyshev collocation and a mapping technique", *SIAM J. Sci. Comput.* (submitted).

M. Dryja and O. B. Widlund (1990), "Towards a unified theory of domain decomposition algorithms for elliptic problems", in Chan et al. (1990).

M. Dubiner (1987), "Asymptotic analysis of spectral methods", *J. Sci. Comput.* 2: 3–31.

M. Dubiner (1991), "Triangular spectral elements", *J. Sci. Comput.* 6: 345–90.

H. Eisen and W. Heinrichs (1992), "A new method of stabilization for singular perturbation problems with spectral methods", *SIAM J. Numer. Anal.* 29: 107–22.

E. Eliasen, B. Machenhauer, and E. Rasmussen (1970), "On a numerical method for integration of the hydrodynamical equations with a spectral representation of the horizontal fields", Report no. 2, Institute of Theoretical Meteorology, University of Copenhagen, Denmark.

B. Engquist and A. Majda (1977), "Absorbing boundary conditions for the numerical simulation of waves", *Math. Comput.* 31: 629–51.

E. Fermi, J. Pasta, and S. Ulam (1966), "Studies of nonlinear problems", Los Alamos Report LA-1940, NM. Reproduced in A. C. Newell, ed. (1974), *Nonlinear Wave Motion,* Amer. Math. Soc., Providence, RI, pp. 143–56.

B. A. Finlayson and L. E. Scriven (1966), "The method of weighted residuals – a review", *Appl. Mech. Rev.* 19: 735–48.

P. F. Fischer and E. M. Rønquist (1994), "Spectral elements for large scale parallel Navier-Stokes calculations", *Comput. Meth. Appl. Mech. Eng.* 116: 69–76.

B. Fornberg (1973), "On the instability of leap-frog and Crank–Nicolson approximations of a nonlinear partial differential equation", *Math. Comput* 27: 45–57.

B. Fornberg (1975), "On a Fourier method for the integration of hyperbolic equations", *SIAM J. Numer. Anal.* 12: 509–28.

B. Fornberg (1977), "A numerical study of 2-D turbulence", *J. Comput. Phys.* 25: 1–31.

B. Fornberg (1981a), "Numerical differentiation of analytic functions", *ACM Trans. Math. Software* 7: 512–26.

B. Fornberg (1981b), "Algorithm 579 CPSC: complex power series coefficients", *ACM Trans. Math. Software* 7: 542–7.

B. Fornberg (1987), "The pseudospectral method: comparisons with finite differences for the elastic wave equation", *Geophysics* 52: 483–501.

B. Fornberg (1988a), "The pseudospectral method: accurate representation of interfaces in elastic wave calculations", *Geophysics* 53: 625–37.

B. Fornberg (1988b), "Generation of finite difference formulas on arbitrarily spaced grids", *Math. Comput.* 51: 699–706.

B. Fornberg (1990a), "High order finite differences and the pseudospectral method on staggered grids", *SIAM J. Numer. Anal.* 27: 904–18.

B. Fornberg (1990b), "An improved pseudospectral method for initial-boundary value problems", *J. Comput. Phys.* 91: 381–97.

B. Fornberg (1992), "Fast generation of weights in finite difference formulas", in *Recent Developments in Numerical Methods and Software for ODEs/DAEs/PDEs* (G. D. Byrne and W. E. Schiesser, eds.), World Scientific, Singapore, pp. 97–123.

B. Fornberg (1995), "A pseudospectral approach for polar and spherical geometries", *SIAM J. Sci. Comput.* (to appear).

B. Fornberg (1996), "Calculation of weights for Hermite-type finite difference schemes", *SIAM J. Sci. Comput.* (submitted).

B. Fornberg and D. M. Sloan (1994), "A review of pseudospectral methods for solving partial differential equations", in *Acta Numerica 1994* (A. Iserles, ed.), Cambridge University Press, pp. 203–67.

B. Fornberg and G. B. Whitham (1978), "A numerical and theoretical study of certain nonlinear wave phenomena", *Phil. Trans. Roy. Soc. London A* 289: 373–404.

J. B. J. Fourier (1822), *Théorie analytique de la chaleur,* Didot, Paris.

D. G. Fox and S. A. Orszag (1973), "Pseudospectral approximation to two-dimensional turbulence", *J. Comput. Phys.* 11: 612–19.

D. Funaro (1987), "A preconditioned matrix for the Chebyshev differencing operator", *SIAM J. Numer. Anal.* 24: 1024–31.

D. Funaro (1992), *Polynomial Approximation of Differential Equations,* Lecture Notes in Physics, vol. m8, Springer-Verlag, Berlin.

D. Funaro and D. Gottlieb (1988), "A new method of imposing boundary conditions in pseudospectral approximations of hyperbolic equations", *Math. Comput.* 51: 599–613.

D. Funaro and O. Kavian (1988), "Approximation of some diffusion evolution equations in unbounded domains by Hermite functions", Report of the Institut Élie Cartan, no. 15, Nancy, France (unpublished).

D. Gaier (1987), *Lectures on Complex Approximation,* Birkhäuser, Boston.

C. S. Gardner, J. M. Greene, M. D. Kruskal, and R. M. Miura (1967), "Method for solving the Korteweg–de Vries equation", *Phys. Rev. Lett.* 19: 1095–7.

C. W. Gear (1971), *Numerical Solution of Ordinary and Partial Differential Equations,* Prentice-Hall, Englewood Cliffs, NJ.

J. A. Glassman (1970), "A generalization of the fast Fourier transform", *IEEE Trans. Comput.* C-19: 105–16.

M. Goldberg and E. Tadmor (1985), "Convenient stability criteria for difference approximations of hyperbolic initial-boundary value problems", *Math. Comput.* 44: 361–77.

G. H. Golub and J. Kautsky (1983), "Calculation of Gauss quadrature with multiple free and fixed knots", *Numer. Math.* 41: 147–63.

D. Gottlieb and L. Lustman (1983), "The spectrum of the Chebyshev collocation operator for the heat equation", *SIAM J. Numer. Anal.* 20: 909–21.

D. Gottlieb, L. Lustman, and E. Tadmor (1987), "Stability analysis of spectral methods for hyperbolic initial-boundary value problems", *SIAM J. Numer. Anal.* 24: 241–58.

D. Gottlieb and S. A. Orszag (1977), *Numerical Analysis of Spectral Methods,* SIAM, Philadelphia.

D. Gottlieb and E. Tadmor (1991), "The CFL condition for spectral approximation to hyperbolic BVPs", *Math. Comput.* 56: 565–88.

D. Gottlieb and E. Turkel (1980), "On time discretization for spectral methods", *Stud. Appl. Math.* 63: 67–86.

P. M. Gresho and R. L. Lee (1981), "Don't suppress the wiggles - they're trying to tell you something", *Computers and Fluids* 9: 223–53.

C. E. Grosch and S. A. Orszag (1985), "Numerical solution of problems in unbounded regions: coordinate transformations", *J. Comput. Phys.* 25: 273–96.

M. M. Gupta (1991), "High accuracy solutions of incompressible Navier–Stokes equations", *J. Comput. Phys.* 93: 343–59.

B. Gustafsson, H.-O. Kreiss, and J. Oliger (1995), *Time Dependent Problems and Difference Methods,* Wiley, New York.

B. Gustafsson, H.-O. Kreiss, and A. Sundström (1972), "Stability theory of difference approximations for mixed initial-boundary value problems II", *Math. Comput.* 26: 649–85.

E. Hairer, S. P. Nørsett, and G. Wanner (1987), *Solving Ordinary Differential Equations I - Non-Stiff Problems,* Springer-Verlag, Berlin.

E. Hairer and G. Wanner (1991), *Solving Ordinary Differential Equations II - Stiff and Differential-Algebraic Problems,* Springer-Verlag, Berlin.

P. Haldenwang, G. Labrosse, S. Abboudi, and M. Deville (1984), "Chebyshev 3-D spectral and 2-D pseudospectral solvers for the Helmholtz equation", *J. Comput. Phys.* 55: 115–28.

L. Halpern and L. N. Trefethen (1988), "Wide-angle one-way wave equations", *J. Acoust. Soc. Amer.* 84: 1397–1404.

D. R. Hartree and J. R. Womersley (1937), "A method for the numerical or mechanical solution of certain types of partial differential equations", *Proc. Royal Soc. London A* 161: 353–66.

J. E. Haugen and B. Machenhauer (1993), "A spectral limited-area model formulation with time-dependent boundary conditions applied to the shallow-water equations", *Monthly Weather Review* 121: 2618–30.

P. Henrici (1964), *Elements of Numerical Analysis,* Wiley, New York.

E. Hewitt and R. E. Hewitt (1979), "The Gibbs–Wilbraham phenomenon: an episode in Fourier analysis", in *History of Exact Sciences,* vol. 21, Springer-Verlag, New York, pp. 129–60.

R. L. Higdon (1990), "Radiation boundary conditions for elastic wave propagation", *SIAM J. Numer. Anal.* 27: 831–70.

R. L. Higdon (1991), "Absorbing boundary conditions for elastic waves", *Geophysics* 56: 231–41.

R. Hirota (1977), "Nonlinear partial difference equations III. Discrete sine-Gordon equation", *J. Phys. Soc. Japan* 43: 2079–86.

E. W. Hobson (1931), *Theory of Spherical and Ellipsoidal Harmonics,* Cambridge University Press.

O. Holberg (1987), "Computational aspects of the choice of operator and sampling interval for numerical differentiation in large-scale simulation of wave phenomena", *Geophys. Prospecting* 35: 629–55.

Y.-C. C. Hou and H.-O. Kreiss (1993), "Comparison of finite difference and the pseudo-spectral approximations for hyperbolic equations" (unpublished).

W. Huang and D. M. Sloan (1993a), "Pole condition for singular problems: the pseudospectral approximation", *J. Comput. Phys.* 107: 254–61.

W. Huang and D. M. Sloan (1993b), "A new pseudospectral method with up-wind features", *IMA J. Numer. Anal.* 13: 413–30.

W. Huang and D. M. Sloan (1994), "The pseudospectral method for solving differential eigenvalue problems", *J. Comput. Phys.* 111: 399–409.

M. Y. Hussaini and T. A. Zang (1984), "Iterative spectral methods and spectral solution to compressible flows", in *Spectral Methods for PDEs* (R. Voigt, D. Gottlieb, and M. Y. Hussaini, eds.), SIAM, Philadelphia.

J. M. Hyman (1979), "A method of lines approach to the numerical solution of conservation laws", in *Advances in Computational Methods for PDEs, III* (R. Vichnevetsky and R. S. Stepleman, eds.), International Association for Mathematics and Computers in Simulation, New Brunswick, NJ, pp. 313–21.

A. Iserles (1986), "Generalized leapfrog methods", *IMA J. Numer. Anal.* 6: 381–92.

A. Iserles and S. P. Nørsett (1984), "A proof of the first Dahlquist barrier by order stars", *BIT* 24: 529–37.

M. Jarraud and A. P. M. Baede (1985), "The use of spectral techniques in numerical weather prediction", in *Large-Scale Computations in Fluid Mechanics,* Lectures in Applied Mathematics, vol. 22, Amer. Math. Soc., Providence, RI.

O. G. Johnson (1984), "Three-dimensional wave equation computations on vector computers", *Proc. IEEE* 72: 90–5.

A. Karageorghis (1988), "A note on the Chebyshev coefficients of the general order derivative of an infinitely differentiable function", *J. Comput. Appl. Math.* 21: 129–32.

G. E. Karniadakis and S. A. Orszag (1993), "Nodes, modes and flow codes", *Physics Today* (March): 32–42.

K. R. Kelly, R. W. Ward, S. Treitel, and R. M. Alford (1976), "Synthetic seismograms: a finite difference approach", *Geophysics* 41: 2–27.

M. Kindelan, A. Kamel, and P. Sguazzero (1990), "On the construction and efficiency of staggered numerical differentiators for the wave equation", *Geophysics* 55: 107–10.

D. A. Kopriva and J. H. Kolias (1995), "A conservative staggered-grid Chebyshev multidomain method for compressible flows", *J. Comput. Phys.* (submitted).

D. Kosloff and E. Baysal (1982), "Forward modeling by the Fourier method", *Geophysics* 47: 1402–12.

D. Kosloff and D. Kessler (1989), "Seismic numerical modeling", in *Geophysical Tomography* (A. Tarantola et al., eds.), North-Holland, Amsterdam, pp. 249–312.

D. Kosloff, M. Reshef, and D. Loewenthal (1984), "Elastic wave calculations by the Fourier method", *Bull. Seism. Soc. Amer.* 74: 875–91.

D. Kosloff and H. Tal-Ezer (1993), "A modified Chebyshev pseudospectral method with an $O(N^{-1})$ time step restriction", *J. Comput. Phys.* 104: 457–69.

R. Kosloff and D. Kosloff (1986), "Absorbing boundaries for wave propagation problems", *J. Comput. Phys.* 63: 363–76.

H.-O. Kreiss (1962), "Über die Stabilitätsdefinition für Differenzengleichungen die partielle Differentialgleichungen approximieren", *Nordisk Tidskr. Informationsbehandling* 2: 153-81.

H.-O. Kreiss (1968), "Stability theory for difference approximations of mixed initial boundary value problems I", *Math. Comput.* 22: 703-14.

H.-O. Kreiss (1970), "Initial boundary value problems for hyperbolic systems", *Comm. Pure Appl. Math.* 23: 277-88.

H.-O. Kreiss and J. Oliger (1972), "Comparison of accurate methods for the integration of hyperbolic equations", *Tellus* 24: 199-215.

H.-O. Kreiss and J. Oliger (1979), "Stability of the Fourier method", *SIAM J. Numer. Anal.* 16: 421-33.

V. I. Krylov (1962), *Approximate Calculation of Integrals*, MacMillan, New York.

J. D. Lambert (1991), *Numerical Methods for Ordinary Differential Systems: the Initial Value Problem*, Wiley, New York.

C. Lanczos (1938), "Trigonometric interpolation of empirical and analytical functions", *J. Math. Phys.* 17: 123-99.

P. Lax (1967), "Hyperbolic difference equations: a review of the Courant-Friedrichs-Lewy paper in the light of recent developments", *IBM Journal* 11: 235-8.

S. K. Lele (1992), "Compact finite difference schemes with spectral-like resolution", *J. Comput. Phys.* 103: 16-42.

I. Lie (1993), "Using implicit ODE methods with iterative linear equation solvers in spectral methods", *SIAM J. Sci. Comput.* 14: 1194-1213.

L. T. Long and J. S. Liow (1990), "A transparent boundary for finite-difference wave simulation", *Geophysics* 55: 201-8.

Y. L. Luke (1969), *The Special Functions and Their Approximations*, Academic Press, New York.

B. Machenhauer (1991), "Spectral methods," in *Numerical Methods on Atmospheric Models*, vol. 1 (seminar proceedings), European Center for Medium Range Weather Forecast, Reading, UK.

R. McLachlan (1994), "The world of symplectic space", *New Scientist* (19 March): 32-5.

Y. Maday, S. M. O. Kaber, and E. Tadmor (1993), "Legendre PS viscosity methods for nonlinear conservation laws", *SIAM J. Numer. Anal.* 30: 321-42.

A. Majda, J. McDonough, and S. Osher (1978), "The Fourier method for nonsmooth initial data", *Math. Comput.* 32: 1041-81.

A. I. Markushevich (1967), *Theory of Functions of a Complex Variable*, vol. III (R. A. Silverman, trans.), Prentice-Hall, Englewood Cliffs, NJ.

C. Mavriplis and J. van Rosendale (1993), "Triangular spectral elements for incompressible fluid flow", ICASE Report 93-100, NASA Langley Research Center, Hampton, VA.

D. F. Mayers (1966), "Convergence of polynomial interpolation", in *Methods of Numerical Approximation* (D. C. Handscomb, ed.), Pergamon Press, Oxford.

B. Mercier (1989), *An Introduction to the Numerical Analysis of Spectral Methods*, Springer-Verlag, Berlin.

W. J. Merryfield and B. Shizgal (1993), "Properties of collocation third-derivative operators", *J. Comput. Phys.* 105: 182-5.

E. Merzbacher (1970), *Quantum Mechanics*, 2nd ed., Wiley, New York.

A. A. Michelson and S. W. Stratton (1898), "A new harmonic analyser", *Phil. Mag.* 45: 85-91.

R. Mittet, O. Holberg, B. Arntsen, and L. Amundsen (1988), "Fast finite difference modeling of 3-D elastic wave equation", *Society of Exploration Geophysics Expanded Abstracts* 1: 1308–11.

L. S. Mulholland and D. M. Sloan (1992), "The role of preconditioning in the solution of evolutionary PDEs by implicit Fourier PS methods", *J. Comput. Appl. Math.* 42: 157–74.

K. L. Nielsen (1956), *Methods in Numerical Analysis,* MacMillan, New York.

F. Z. Nouri and D. M. Sloan (1989), "A comparison of Fourier pseudospectral methods for the solution of the Korteweg–de Vries equation", *J. Comput. Phys.* 83: 324–44.

G. G. O'Brien, M. A. Hyman, and S. Kaplan (1951), "A study of the numerical solution of partial differential equations", *J. Math. Phys.* 29: 223–51.

L. Onsager (1949), "Statistical hydrodynamics", *Del Nuovo Cimento* N2 (Suppl. Al, vol. VI, ser. IX): 279–87.

S. A. Orszag (1969), "Numerical methods for the simulation of turbulence", *Phys. Fluids Suppl. II* 12: 250–7.

S. A. Orszag (1970), "Transform method for calculation of vector coupled sums: Application to the spectral form of the vorticity equation", *J. Atmosph. Sci.* 27: 890–5.

S. A. Orszag (1972), "Comparison of pseudospectral and spectral approximations", *Stud. Appl. Math.* 51: 253–9.

S. A. Orszag (1974), "Fourier series on spheres", *Monthly Weather Review* 102: 56–75.

S. A. Orszag (1980), "Spectral methods for problems in complex geometries", *J. Comput. Phys.* 37: 70–92.

S. A. Orszag and L. C. Kells (1980), "Transition to turbulence in plane Poiseuille flow and plane Couette flow", *J. Fluid Mech.* 96: 159–205.

A. T. Patera (1984), "A spectral element method for fluid dynamics: laminar flow in a channel expansion", *J. Comput. Phys.* 54: 468–88.

D. Pathria (1994), "The correct formulation of intermediate boundary conditions for Runge–Kutta time integration of initial-boundary value problems" (in preparation).

N. A. Phillips (1959), "An example of nonlinear computational instability", in *The Atmosphere and the Sea in Motion,* Rockefeller Inst. Press, New York, pp. 501–4.

T. N. Phillips (1988), "On the Legendre coefficients of the general-order derivative of an infinitely differentiable function", *IMA J. Numer. Anal.* 8: 455–9.

T. N. Phillips, T. A. Zang, and M. Y. Hussaini (1986), "Preconditioners for the spectral multigrid method", *IMA J. Numer. Anal.* 6: 273–92.

M. J. D. Powell (1981), *Approximation Theory and Methods,* Cambridge University Press.

S. C. Reddy and L. N. Trefethen (1990), "Lax-stability of fully discrete spectral methods via stability regions and pseudo-eigenvalues", *Comp. Meth. Appl. Mech. Engr.* 80: 147–64.

R. Renaut and J. Frölich (1995), "A pseudospectral Chebyshev method for the 2-D wave equation with domain stretching and absorbing boundary conditions", *J. Comput. Phys.* (submitted).

L. F. Richardson (1910), "The approximate arithmetical solution by finite differences of physical problems involving differential equations, with an application to the stresses in a masonry dam", *Phil. Trans. Royal Soc. London* 210: 307–57.

L. F. Richardson (1922), *Weather Prediction by Numerical Process,* Cambridge University Press.

R. D. Richtmyer and K. W. Morton (1967), *Difference Methods for Initial-Value Problems,* 2nd ed., Wiley, New York.

T. J. Rivlin (1969), *An Introduction to the Approximation of Functions,* Dover, New York.

P. Roe (1994), "Linear bicharacteristic schemes without dissipation", ICASE Report 94-65, NASA Langley Research Center, Hampton, VA.

P. Rosenau and J. M. Hyman (1993), "Compacton: soliton with finite wavelength", *Phys. Rev. Lett.* 70: 564-7.

J. Sand and O. Østerby (1979), "Regions of absolute stability", Report DAIMI PP-102, Computer Science Department, Aarhus University, Denmark.

G. Sansone (1959), *Orthogonal Functions,* Interscience, New York.

J. M. Sanz-Serna and M. P. Calvo (1994), *Numerical Hamiltonian Problems,* Chapman & Hall, London.

A. Schönhage (1961), "Fehlerfortpflanzung bei Interpolation", *Numer. Math.* 3: 62-71.

L. F. Shampine and M. K. Gordon (1975), *Computer Solution of Ordinary Differential Equations,* Freeman, San Francisco.

A. Solomonoff (1992), "A fast algorithm for spectral differentiation", *J. Comput. Phys.* 98: 174-7.

A. Solomonoff (1994), "Bayes finite difference schemes", *SIAM J. Numer. Anal.* (submitted).

A. Solomonoff and E. Turkel (1989), "Global properties of pseudospectral methods, *J. Comput. Phys.* 81: 239-76.

F. Stenger (1993), *Numerical Methods Based on Sinc and Analytic Functions,* Springer Series in Computational Mathematics, vol. 20, Springer-Verlag, Berlin.

G. Strang (1986), *Introduction to Applied Mathematics,* Wellesley-Cambridge Press, Wellesley, MA.

A. Stuart (1989), "Nonlinear instability in dissipative finite difference schemes", *SIAM Review* 31: 191-220.

P. N. Swarztrauber (1974), "The direct solution of the discrete Poisson equation on the surface of a sphere", *J. Comput. Phys.* 15: 46-54.

G. Szegö (1959), *Orthogonal Polynomials,* Amer. Math. Soc., Providence, RI.

E. Tadmor (1987), "Stability analysis of finite-difference, pseudospectral and Fourier-Galerkin approximations for time-dependent problems", *SIAM Review* 29: 525-55.

E. Tadmor (1989), "Convergence of spectral methods for nonlinear conservation laws", *SIAM J. Numer. Anal.* 26: 30-44.

E. Tadmor (1990), "Shock capturing by the spectral viscosity method", *Comp. Meth. in Appl. Mech. Eng.* 80: 197-208.

E. Tadmor (1993), "Superviscosity and spectral approximations of nonlinear conservation laws", in *Numerical Methods for Fluid Dynamics IV* (M. J. Baines and K. W. Morton, eds.), Oxford University Press, pp. 69-82.

T. R. Taha and M. J. Ablowitz (1984), "Analytical and numerical aspects of certain nonlinear evolution equations. III. Numerical, Korteweg-de Vries equation", *J. Comput. Phys.* 55: 231-53.

H. Tal-Ezer, J. M. Carcione, and D. Kosloff (1990), "An accurate and efficient scheme for wave propagation in linear viscoelastic media", *Geophysics* 55: 1366-79.

T. Tang (1993), "Hermite spectral method for Gaussian type functions", *SIAM J. Sci. Comput.* 14: 594-606.

T. Tang and M. R. Trummer (1994), "Boundary layer resolving pseudospectral methods for singular perturbation problems", *SIAM J. Sci. Comput.* (submitted).

F. Tappert (1974), "Numerical solutions of the Korteweg-de Vries equation and its generalizations by the split-step Fourier method", in *Lectures in Applied Mathematics,* vol. 15, Amer. Math. Soc., Providence, RI, pp. 215–16.

T. D. Taylor, R. S. Hirsh, and N. M. Nadworny (1984), "Comparison of FFT, direct inversion and conjugate gradient methods for use in pseudospectral methods", *Computers and Fluids* 12: 1–9.

E. Tessmer and D. Kosloff (1994), "3-D elastic modeling with surface topography by a Chebyshev spectral method", *Geophysics* 59: 464–73.

M. Thuné (1990), "A numerical algorithm for stability analysis of difference methods for hyperbolic systems", *SIAM J. Sci. Comput.* 11: 63–81.

P. A. Tirkas, C. A. Balanis, and R. A. Renaut (1992), "High order absorbing boundary conditions for the finite-difference time-domain method", *IEEE Trans. Antennas and Propagation* 40: 1215–22.

L. N. Trefethen (1982), "Group velocity in finite difference schemes", *SIAM Review* 24: 113–36.

L. N. Trefethen (1983), "Group velocity interpretation of the stability theory of Gustafsson, Kreiss and Sundström", *J. Comput. Phys.* 49: 199–217.

L. N. Trefethen (1988), "Lax-stability vs. eigenvalue stability of spectral methods", in *Numerical Methods for Fluid Dynamics III* (K. W. Morton and M. J. Baines, eds.), Clarendon Press, Oxford, pp. 237–53.

L. N. Trefethen, A. E. Trefethen, S. C. Reddy, and T. A. Driscoll (1993), "Hydrodynamic stability without eigenvalues", *Science* 261: 578–84.

L. N. Trefethen and M. R. Trummer (1987), "An instability phenomenon in spectral methods", *SIAM J. Numer. Anal.* 24: 1008–23.

L. N. Trefethen and J. A. C. Weideman (1991), "Two results on polynomial interpolation in equally spaced points", *J. Approx. Theory* 65: 247–60.

A. H. Turetskii (1940), "The bounding of polynomials prescribed at equally distributed points" (Russian), *Proc. Pedag. Inst. Vitebsk.* 3: 117–27.

A. H. Turetskii (1968), *Theory of Interpolation in Problem Form* (Russian), Vyshóeìish Shkola, Minsk.

C. van Loan (1992), *Computational Frameworks for the Fast Fourier Transform,* SIAM, Philadelphia.

P. Vértesi (1990), "Optimal Lebesgue Constant for Lagrange Interpolation", *SIAM J. Numer. Anal.* 27: 1322–31.

A. C. Vliegenthart (1971), "On finite-difference methods for the Korteweg-de Vries equation", *J. Eng. Math.* 5: 137–55.

R. G. Voigt, D. Gottlieb, and M. Y. Hussaini, eds. (1984), *Spectral Methods for Partial Differential Equations,* SIAM, Philadelphia.

J. L. Walsh (1960), *Interpolation and Approximation by Rational Functions in the Complex Domain,* Colloquium Publications of the AMS, vol. 20, 3rd ed., Amer. Math. Soc., Providence, RI.

G. Wanner, E. Hairer, and S. P. Nørsett (1978), "Order stars and stability theorems", *BIT* 18: 475–89.

J. A. C. Weideman (1992), "The eigenvalues of Hermite and rational spectral DMs", *Numer. Math.* 61: 409–31.

J. A. C. Weideman and L. N. Trefethen (1988), "The eigenvalues of second-order spectral differentiation matrices", *SIAM J. Numer. Anal.* 25: 1279–98.

B. D. Welfert (1994), "A remark on pseudospectral differentiation matrices", *SIAM J. Numer. Anal.* (submitted).

G. B. Whitham (1974), *Linear and Nonlinear Waves,* Wiley, New York.

E. T. Whittaker (1915), "On the functions which are represented by the expansions of the interpolation theory", *Proc. Roy. Soc. Edinburgh* 35: 181–94.

J. M. Whittaker (1927), "On the cardinal function of interpolation theory", *Proc. Edinburgh Math. Soc.* (1) 2: 41–6.

H. Wilbraham (1848), "On a certain periodic function", *Cambridge and Dublin Mathematical Journal* 3: 198–201.

S. Winograd (1978), "On computing the discrete Fourier transform", *Math. Comput.* 32: 175–99.

D. C. Witte and P. G. Richards (1990), "The pseudospectral method for simulating wave propagation", *Computational Acoustics* 3: 1–18.

H. B. Yao, D. G. Dritschel, and N. J. Zabusky (1994), "High gradient phenomena in 2-D vortex interactions", *Physics of Fluids A* (submitted).

S. Y. K. Yee (1981), "Solution of Poisson's equation on a sphere by truncated double Fourier series", *Monthly Weather Review* 109: 501–5.

H. Yoshida (1990), "Construction of higher order symplectic integrators", *Phys. Lett. A* 150: 262–8.

N. J. Zabusky and M. D. Kruskal (1965), "Interactions of 'solitons' in a collisioniess plasma and the recurrence of initial states", *Phys. Rev. Lett.* 15: 240–3.

Index

Printed in the United States
By Bookmasters